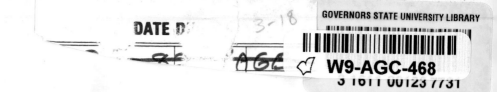

GEOMETRY FOR ELEMENTARY TEACHERS

GEOMETRY FOR ELEMENTARY TEACHERS

JOHN E. YOUNG

Southeast Missouri State College

GRACE A. BUSH

Kent State University

HOLDEN-DAY

San Francisco, Cambridge, London, Amsterdam

ISBN 0-8162-9984-6

Library of Congress Catalog Card Number: 77-155559

Printed in the United States of America.

to
   Bryan
   Elizabeth
   Eric
   Kirsten
   Larry
   Melinda
   Rebecca

PREFACE

Geometry, as a course for elementary education majors, has appeared on the scene rather recently, owing largely to the pressures of society and the urging of the Committee on the Undergraduate Program in Mathematics (CUPM).

The needs of elementary education majors in the area of mathematics are special indeed. These students are generally good students, interested in education, whose motivation in mathematics, in general, falls short of that of students in the sciences. Yet these elementary education majors are expected to teach mathematics, including geometry, to impressionable youngsters. The authors feel that a rigorous mathematical development of geometry would be wrong for this audience.

So we have put together in this book a blend of mathematics as found in (1) everyday experiences, (2) special classroom experiments, and (3) formal Euclidean geometry. We believe this has been done without sacrificing a great deal of mathematical purity; how well we have succeeded is left to the reader. We do know this material works in the classroom with the elementary teachers we have faced, and that at the conclusion of the course most of the students have indicated an understanding of geometric principles.

The primary difference between this text and others is our emphasis on experience and experimentation, through which we have tried to lead to sound mathematical reasoning while retaining mathematical precision. Any deviation, unless noted, is unintentional.

The student does not get bogged down in rigor and proof but is constantly reminded that a given idea relates to experience. The student is asked to think about experiences to see what conjectures can be developed. We remind the reader that the final audience for this book is the elementary school pupil, and it is to this end that we write.

The authors assume the reader has had previous mathematics courses involving sets and logical deduction. However, for those students whose memories need jogging, or who had such material long ago, Chapter 1 reviews the foundations. It contains as much set algebra as is needed for the rest of the book, in which geometry is treated from a *set of points* approach. A discussion of deductive and inductive reasoning is included to ensure understanding on the part of the student of these two very important approaches to a new mathematics problem. This chapter can be omitted for a short course, or where the instructor feels the students are well grounded.

Chapters 2 through 4 could be termed "geometry without measure" and should be taught consecutively without omission. Mathematical precision is maintained, except in a few instances where an idea is beyond the scope of the course (such as general congruence or congruence of curves). One feature in these chapters is the integration of many of the classic Euclidean constructions into the experiments and exercises instead of presentation as separate units. This avoids giving them the appearance of a set of separate ideas.

Chapters 5 through 7 develop the idea of measure in geometry and present the usual geometric formulas for plane and solid figures as well as the ingredients of the English and metric systems of measure. Topics relating to the accuracy and dependability of measure are given in Chapter 7; something often overlooked by this kind of audience. The authors intend, by this order of topics, that the reader understand that geometry is a study of properties of geometry and is not dependent on measure, or, alternatively, can be measured in several ways. Too often, geometry means measure to the elementary school teacher. This seems to be a block that our traditional educational system has built into it.

Part of Chapter 4 and all of Chapter 8 are a re-examination of topics covered in other sections using the ideas of transformations and analytic geometry. Although Chapter 8 can be omitted for a shorter course, the sections on transformations should be included in the content of the basic course.

The authors find more than enough material here for a 3-semester-hour course. Chapters 2 through 7 are sufficient for a 3-semester-hour course with 45 50-minute lectures and a suitable number of testing periods. A more advanced class might cover Chapters 2 through 8 as an alternative. A 4- or 5-semester-hour course could include all eight chapters.

The authors consider themselves responsible for all errors and take full responsibility for them. We wish to thank our respective colleagues and spouses for the endurance and patience that is required when associated with a writing project, while not being directly involved. Special thanks go to our typist, Mrs. H. N. Abhau; to Professor Byron Dressler, Director of the Computer Center at Kent State University; and to Professor R. J. Michel, Chairman of the Mathematics Department at Southeast Missouri State College, for his cooperation in arranging classes for the use of these materials.

Cape Girardeau, Missouri   *John E. Young*

Kent, Ohio   *Grace A. Bush*

August, 1970

TABLE OF CONTENTS

GEOMETRY FOR ELEMENTARY TEACHERS

1 / INTRODUCTION TO GEOMETRY

1.1 HISTORICAL MOTIVATION

You may recall high-school geometry as a course which dealt primarily with lines, triangles, axioms, and proofs, with little relationship to the real world. However, long before you studied geometry, you were introduced to geometric ideas. Even in pre-school years children can identify basic shapes such as squares, circles, and triangles and can understand the difference between straight and curved lines. Many youngsters who have never heard the word *congruent* have grasped the *concept* of congruence in, say, working jigsaw puzzles where it is necessary to find a piece to "fit" exactly into a given space, or in so simple a task as cutting a piece of string to the exact length of another piece of string. The fact that special shapes, such as the letters making up the words on this page, convey meaning demands a theoretically congruent figure in reproduction if the meaning is to remain. Mechanical typesetting processes guarantee that all reproductions of a lowercase "a" are enough like each other, and enough different from other letters, to avoid confusion in reading. By coloring pictures, children gain an intuitive idea of the separation of a plane surface, that is, the division of the parts of the page into sections to be colored differently. The typical classroom seating arrangement is a good example of rows and columns, the beginnings of co-ordinate geometry. These geometric relationships are not always pointed out in the classroom but they do exist, and students develop many intuitive geometric ideas simply by being exposed to them.

Studies recently have shown that geometry can and should be included in the grade-school curriculum. Children like to work in "visual

1

geometry," and they enjoy discovering geometric properties through the manipulation of geometric models. In the lower grades geometry is usually a study of basic figures, division of space, components of shapes, and real applications. Because we live in three dimensions, the study of three-dimensional geometry is included.

If it is agreed that children are to study geometry in the grades and that they are to discover for themselves the fundamental geometric properties, it is necessary to have teachers to suggest experiments for the students to undertake, to guide the students in their discoveries, and to help the students formulate conclusions from their experiments. In this course you will be introduced to many ideas in geometry. You will be led to the discovery of many relationships in geometry as you draw conclusions from your own experiments. When it seems appropriate, a traditional deductive proof of the properties will also be included. This text is concerned with an orderly development of geometric properties and their applications. When you have completed your study of the material in this text you should be prepared to guide grade-school students in their development of geometric properties.

1.2 TWO APPROACHES TO GEOMETRY

Geometry can be developed as a mathematical study in two different ways. These two approaches are usually called the *inductive* and *deductive* methods of development. Rather than oppose each other they should supplement one another in an ideal learning situation. Both methods will be employed in this text.

Historically, the development of geometry was started because of need. Once mankind settled into an agrarian form of life where establishment of boundary lines, subdivision of property, and construction of permanent structures was a must, then the need for a workable and useful geometry became imperative.

Doubtless, the first geometric "facts" were found by experimentation and observation. Historians credit the Egyptians with knowing that a triangle with sides of 3, 4, and 5 units was a right triangle. The rope stretchers of 4000 years ago used this knowledge to devise a portable means of determining right angles. Many other relationships, some correct, others inaccurate, were determined experimentally by this same civilization. These relationships made up a practical or working geometry. The word *geometry* indicates its practical beginning; the literal meaning of the word is "earth-measure."

We know that the Egyptians were astute observers of three-dimensional geometric relationships. Some of their constructions, such as the

2

pyramids, are amazingly accurate models of geometric shapes. The ratio of the diameter of a circle to its circumference was particularly troublesome to the ancient geometers, and it was not until recent times that the value of this ratio, denoted π, could be approximated with any desired degree of accuracy. The Egyptians, according to the Rhind Papyrus,[1] used $(8d/9)^2$ to find the area of the circle with diameter d. This gives a value of $\pi = 3.16$ (to two decimal places); an approximation which doubtless was obtained experimentally. The Babylonians of the pre-Christian era developed much of the geometry of planetary movements and were able to predict future positions of the heavenly bodies. Their predictions must have been based on accurate observation of, and some insight into, the three-dimensional nature of the solar system.

We can suppose that some deductive thinking was applied by these two ancient societies, but there is no evidence of deduction in mathematics until after the rise of Greek civilization (beginning about 1000 B.C.). The Greeks, believing in absolutes, assumed absolute truths or rules in mathematics as they did in all disciplines. One of these is a rule of inference which is illustrated in the following argument.

If you accept as true the statements, "All books are made of paper" and "All paper is flammable," then logically you must also accept the conclusion, "All books are flammable."

That is, the conclusion "follows logically" from the given statements. Another example of the same kind of reasoning is the following: If we agree that

All polygons have interior angles

and

All triangles are polygons,

then we must conclude that

All triangles have interior angles.

Although the statements about books and triangles are reasonable, logic can be applied to nonsense statements or statements concerning things about which we know little or nothing. If you are willing to accept the statements

All windfops are smarteps

and

All smarteps are gleepers,

[1] J. R. Newman, *The World of Mathematics*, Simon & Schuster, New York, 1956, vol. 1, pp. 174–175.

3

then you must agree that

All windfops are gleepers

is a logical consequence of the first two statements.

Geometry was developed by the Greeks on such reasoning processes. As a direct result of this development, students in a traditional high-school course in geometry are asked to accept without proof certain basic properties called *axioms* or *postulates*. Some of these are:

The whole is equal to the sum of its parts.

A quantity is equal to itself. (*reflexive property of equality*)

If equals are subtracted from equals the remainders are equal. (*subtraction property of equality*)

Two quantities equal to the same quantity are equal to each other. (*transitive property of equality*)

The reason given for accepting these statements is that they are "self-evident." With several undefined terms, a set of definitions, and such a set of *axioms* or *postulates*, the course begins. New truths or theorems are established from the definitions and axioms. Once a statement is established as true it can be used in establishing the truth of other theorems or conjectures. If the original set of truths includes the axioms above and the definition, "A *straight angle*, that is, a straight line, has a measure of 180

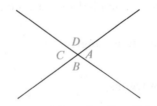

ILLUSTRATION 1.1 $m(\sphericalangle A) = m(\sphericalangle C)$

degrees," then the following statements can be made about the angles in Illus. 1.1. [The measure of an angle A is denoted here by $m(\sphericalangle A)$.]

$m(\sphericalangle A) + m(\sphericalangle B) = 180°$	Since these two angles form a straight angle
$m(\sphericalangle B) + m(\sphericalangle C) = 180°$	Since these two angles form a straight angle
$m(\sphericalangle A) + m(\sphericalangle B) = m(\sphericalangle B) + m(\sphericalangle C)$	Transitive property of equality

4

$$m(\measuredangle A) = m(\measuredangle C)$$ Subtraction property of equality

This last statement is usually given in the form, "Vertical angles have equal measure." At the moment we are not concerned with angle measure; the important notion for us now is: If you accept the axioms given earlier, then you must also accept the statement

Vertical angles have equal measure.

This is an example of the *axiomatic* or direct method of proof using deductive reasoning: Only definitions and accepted truths are used in the proofs. You may question the assumed statements, but once these statements are agreed on, they may be used to develop further ideas. Each proof follows logically from previous statements, the axioms, and definitions.

This is an artificial presentation however, and *not at all the manner in which geometry was developed.* With the hindsight we now have it is relatively easy to decide which statements should serve as axioms and the order in which the rest of the truths should be presented so that the development is completely logical. Pioneers in geometry did not have the advantage of a pool of information to look back to and used quite different methods.

The Egyptians were well acquainted with triangles, circles, and squares, knew how to find areas and volumes, and applied their knowledge to build magnificent structures with great accuracy. It was mentioned earlier that they knew that if the sides of a triangle had the 3-4-5 relationship, then one of its angles was a right angle; they had found this property purely through experimentation. They could see by marking squares on each side of a right triangle with legs of 3 units and 4 units, respectively, that (see Illus. 1.2) $3^2 + 4^2 = 5^2$. Similarly, by laying out tiles and counting, they could see that $5^2 + 12^2 = 13^2$. Their limited understanding of the number system stopped them from going further, and it was not until much later that it was shown deductively (by Pythagoras) that in *every* right triangle the sum of the squares of the two shorter sides equals the square of the *longest side* or *hypotenuse*.

We could use an approach similar to that of the Egyptians to compare the vertical angles formed by the crossed lines in Illus. 1.1. Using a protractor, we would measure the angle at A and the angle at C. Our two measurements might or might not be the same, as it is quite likely there would be some error. But they would probably be quite close in value, and we would conclude that angle A and angle C probably had the

5

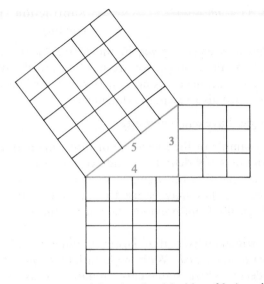

ILLUSTRATION 1.2   A right triangle with sides of 3, 4, and 5 units

same measure. If we acted as the Egyptians, we would be interested only in this pair of vertical angles and would not extend our results to other vertical angles. However, if we drew several more pairs of intersecting lines and measured vertical angles in each pair, we would soon see that in every case the opposite or vertical angles appeared to have equal measure. We might then conclude that vertical angles always have equal measure. This is an *extension of empirical evidence to cases yet to be examined.* Arriving at a conclusion in this manner is "inducing" a property; thus the name *inductive* reasoning.

Although in this case the conclusion is correct, it is necessary to issue a word of caution. *It is never possible to prove that something is always true simply by verifying that it is true in a few (or even many) cases.* A conclusion derived by experience alone is always suspect. One reason why the laws of physics and other sciences are constantly being revised is because they are developed experimentally; new instruments yield new data which do not fit old theories. A mathematical example of erroneous inductive reasoning will illustrate the dangers of using this method.

In mathematics, a prime number is a positive integer greater than 1 which is evenly divisible by only two positive integers, the integer 1 and itself. Examples of prime numbers are 2 and 7, for each of these numbers can be evenly divided by itself and by 1 but by no other positive integer. Since 4 can be divided evenly by 1, 2, and 4, it is not a prime.

6

Suppose that we somehow decided that the sum of two prime numbers is always a prime number. To prove that this is true we show that

$$2 + 3 = 5$$
$$2 + 17 = 19$$
$$2 + 29 = 31$$
$$2 + 41 = 43$$

This relatively short list can be extended to a much longer list; each entry, displaying a set of two prime numbers whose sum is also a prime number, would be a piece of evidence for our conjecture. However, you can say, "But 3 + 5 = 8, and 8 is not a prime," and with just one illustration you have shown that the conjecture is not true. Thus, the weakness of inductive reasoning is that you may obtain false conclusions on insufficient evidence. We based the conclusion on several examples and can invalidate it with one counterexample. In the case of the statement, "Vertical angles have equal measure," each measurement reinforces the original conjecture, but one counterexample, if it exists, would destroy it. We need another, better, means of obtaining the same conclusion.

Both direct or deductive reasoning and experimental or inductive reasoning have been used in geometry for many years. In the example concerning vertical angles, deductive reasoning definitely established the conjecture in just a few steps with no room for doubt, once the premises were accepted. The same conclusion can be reached using inductive methods, but then there would always be the possibility of the result being destroyed by a single counterexample.

The deductive proof of this vertical-angle theorem was simplified greatly by the statement, "A straight angle has a measure of 180 degrees," which was accepted as a definition. However, this may not be necessary. If we could prove that a straight angle must have a measure of 180 degrees, then we could delete this item from the "must be accepted as true" list. Similarly, if we could prove the transitive property of equality, we could remove that from the list. Generally, the list of things which must be accepted as true without proof should be made as short as possible.

It should be noted at this point that inductive reasoning is a powerful and useful tool. Few conclusions would be made in mathematics or any other area without inductive reasoning to help show the way, but we should like to depend on deductive processes as a basis for our conclusions. In this text we shall show many geometric relationships and develop them inductively; then we shall present and discuss the deductive structure.

Naturally, the authors have seen the properties in question many

7

times and know what can and cannot be deduced. The novice cannot expect to guess right every time, or induce correct concepts consistently: deductive geometry was certainly not established without errors in reasoning. Opportunities for experimentation and then verification of experimental results will be provided in the exercises.

EXPERIMENT 1A

Perform each of the following experiments three times; then see if you can make a conjecture about the results.

1 Add several consecutive odd natural numbers beginning with 1 and compare the sum in each case with the number of addends. When is the sum odd? When is it even?

2 Add any three consecutive natural numbers together and compare the sums. When do you get an even sum? An odd one?

3 Make a conjecture relating the sum to the number of addends in each of the following:

$$1 = 1$$
$$1 + 3 = 4$$
$$1 + 3 + 5 = 9$$
$$1 + 3 + 5 + 7 = 16$$
$$1 + 3 + 5 + 7 + 9 = 25$$

4 In the figures, connect the given points in order $(A, B, C, \ldots)$. How many lines did you draw in each case? How can you relate the number of lines drawn to the number of points?

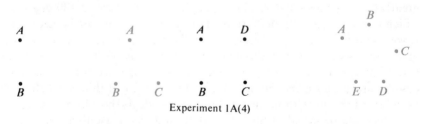

Experiment 1A(4)

5 In the figures for Experiment 1A(4), connect pairs of points in all possible ways. How many lines did you draw in each case? Relate the number of lines to the number of points.

6 Cut each of the figures shown into triangular regions by drawing line segments from A to each of the other lettered points. Relate the number of triangles to the number of labelled points.

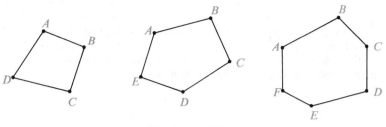

Experiment 1A(6)

7 Construct five triangles of different sizes and shapes. Select one of the angles of one of the triangles and place one of its sides on a horizontal line on paper. Draw a line along the other side of the angle from the vertex as shown. Turn the triangle so one side of another angle is along the line just drawn, and draw a second line to indicate the terminal side of the second angle. Repeat for the third angle. What is the sum $m(\angle A) + m(\angle B) + m(\angle C)$? Repeat this experiment for the other triangles. Is the result always the same? Can you make a conjecture?

Experiment 1A(7)

8 Cut two triangles of exactly the same shape from light cardboard or heavy paper. See if you can fit them together to form a new geometric shape. Repeat for a different set of triangles. What do the new shapes have in common?

9 Each of the triangles shown is drawn with two equal angles. Compare the lengths of the sides opposite these angles by measuring with a ruler. Are the sides equal? Would you conclude this is always the case? How many triangles would you have to draw to be positive that if two angles are equal then two sides are also equal?

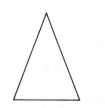
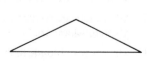

Experiment 1A(9)

9

10 Cut a circle from light cardboard. Cut a second circle with a diameter twice as large. Put a small mark on the edge of each circle, and place the circles on a flat surface with marks aligned. Holding the large circle steady, roll the little circle completely around the large one. How many times did the mark on the smaller circle touch the big circle? Repeat for two other circles. How many revolutions did the smaller circle make?

EXERCISE 1A

1 List two conclusions you have come to concerning each of the pairs of items below:
(a) The quality of a food and the length of time it is cooked
(b) The type of body on an automobile and the age of the driver
(c) The quality of an article of clothing and its cost
(d) The number of rooms in a house and the number of inhabitants
(e) The intelligence of people and their incomes
Can you see anything risky about such inductive conclusions? Do you think most people would agree with your conclusions?

Write the logical conclusion, if one exists, for each of the pairs of statements in Exercises 1A(2) to 1A(11).

2 All freshmen must take mathematics.
John Reed is a freshman.

3 All sophomores are interesting.
All high-school students are interesting.

4 All three-sided closed figures are triangles.
All triangles are polygons.

5 All squares have four sides.
All four-sided figures are quadrilaterals.

6 Some triangles have a right angle.
Some figures with right angles are squares.

7 All intelligent students can pass mathematics.
Helen can pass mathematics.

8 All ostriches are timid.
Victor is timid.

9 No college students are lazy.
All sophomores are college students.

10 All farmers are residents of the United States.
All Nebraskans are farmers.

11 All girls with blonde hair are beautiful.
Nancy does not have blonde hair.

10

12 In each of Exercises 1A(2) to 1A(11) decide if the given statements are true. Can you arrive at an *untrue* conclusion with valid reasoning if one of the assumptions is untrue? Can you arrive at a *true* conclusion with valid reasoning if one of the assumptions is untrue? Illustrate your answers.

13 Using the conclusions you developed inductively in Experiments 1A(6) and 1A(7) as premises, form a logical deductive conclusion about the sum of the interior angles of any polygon.

14 In Experiment 1A(9), if the little circle had been rotated around the inside of the big circle rather than the outside would your answers be the same? Try it.

15 A boy flips a coin five times and gets heads five times. What is his inductive conclusion? Can you state premises to arrive at a deductive conclusion?

1.3 SETS

One of the fundamental concepts in mathematics is that of a set. In this and succeeding sections we shall consider sets, manipulations on sets, and set notation. Since the geometry in this text is based on the set concept, the ideas developed here will be used frequently in later sections.

The term *set* is undefined but a set is thought of as a *collection* or group of distinct elements specified in one of two ways—either by listing the members of the collection or by stating a property common to the members of the set. Some examples of sets are:

    (1)   The set of even natural numbers

    (2)   The set of points on a line

    (3)   The set whose elements are a, e, i, o, u

    (4)   The set of natural numbers which can be substituted for x so that $x + 5 = 8$ is a true statement

    (5)   The set of vertices of a rectangle

    (6)   The set of lines passing through a given point

Example (3) is the only example in which members are listed. The other examples supply rules for determining whether or not a given element belongs to a set.

Throughout our discussion we shall assume the following about sets:

    (1)   *A set must be well defined.* It is necessary that the rule for determining membership in the set makes clear whether an element is in the set. The set of *pretty pictures* is not well defined, as no criterion is given for deciding what *pretty* means. The set of lines through point p is well defined.

    (2)   *The elements in a set are distinct.* An element listed more than once is the same as an element listed once. The set whose ele-

11

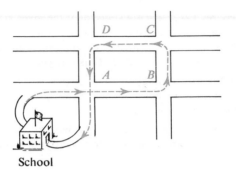

School

ILLUSTRATION 1.3   The set of intersections contains A, B, C, and D

ments are A, B, B, C, C, C is the same as the set whose elements are A, B, C, while the set of street intersections crossed by a school bus on its route (Illus. 1.3) would be denoted by just A, B, C, D, even though it passes corner A twice.

(3) *The order of the elements in the set is not significant.*   The set whose elements are a, b, c is exactly the same as the set containing b, a, c.

1.4   SET NOTATION

Several symbols can be used to denote sets; the most common of these is braces { }. To denote the set consisting of doll, chair, and newspaper we write

$$\{\text{doll, chair, newspaper}\}$$

Similarly, the set of positive integers less than 5 can be written

$$\{+1, +2, +3, +4\}$$

This same set can be denoted by

$$\{w \mid w \text{ is a positive integer less than 5}\}$$

which is read "the set of all w such that w is a positive integer less than 5." The first notation *listed* the elements in the set whereas the second *described* the elements of the set.

Capital letters are ordinarily used to name sets. For example C may represent the set of all closed curves, and L the set of all straight lines. Lowercase letters are used to denote the elements of a set. For example, e might represent an ellipse in C or b a particular line in L. A shorthand notation for stating that e is an element of C is $e \in C$, where the symbol

12

$\in$ is read "is an element of." The symbol $\notin$ is read "is not an element of."

1.5 SUBSETS

Consider the set $S = \{1, 2, 3, 4, 5\}$. The set $A = \{1, 2, 3\}$ consists only of elements selected from S. There are many other such sets, for example, $B = \{1, 3, 5\}$ and $C = \{3\}$. A, B, and C are called *subsets* of S.

Definition 1.1   The set A is a *subset* of set B if and only if every element of A is an element of B. We denote this by $A \subseteq B$, and read, "A is a subset of B."

By this definition the set $K = \{1, 2, 3, 4, 5\}$ is a subset of $S = \{1, 2, 3, 4, 5\}$. Here K contains *all* the elements of S, but the definition does not disallow this. To differentiate between the subset that contains all the elements of a given set and subsets that do not, we state a second definition.

Definition 1.2   The set A is a *proper subset* of the set B if every element of A is an element of B but not every element of B is an element of A. We denote this by $A \subset B$, read, "A is a proper subset of B."

Three proper subsets of $\{a, b, c\}$ are $\{a, b\}$, $\{a, c\}$, and $\{b\}$. Can you name others? Similarly, a proper subset of $\{@, \%, \#, *\}$ is $\{*, \%\}$. Can you name others?

Definition 1.3   The set containing the totality of elements for any particular discussion is called the *universal set* and is denoted by U.

The universal set will vary from discussion to discussion. For a discussion concerning the sets $A = \{1, 2, 3\}$ and $B = \{7, 8, 9\}$, one possible universal set is $U_1 = \{1, 2, 3, 7, 8, 9\}$. Similarly the set $U = \{1, 2, 3, 4, 5, 6, 7, 8, 9\}$ is a universal set for the sets A and B as it too contains all the elements in the discussion (and some others). A universal set for the two sets $L = \{s \mid s$ is a straight line$\}$ and $C = \{c \mid c$ is a curved line$\}$ is the set of all lines. A universal set which includes the set of all right triangles and the set of all equal-sided triangles is the set of all triangles. Another universal set which includes these two sets of triangles as subsets is the set of all three-, four-, and five-sided plane figures.

For the universal set *all triangles*, the subsets *right triangles* and *triangles with exactly two equal sides* are depicted in Illus. 1.4. This diagram indicates that some right triangles (but not all) have exactly two equal sides and that some triangles with exactly two equal sides are right triangles. The unoccupied space in the interior of the area designated *all*

13

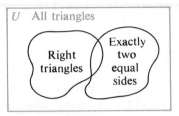

ILLUSTRATION 1.4   The universal set *all triangles* and some of its subsets

triangles is reserved for types of triangles not included in this discussion, that is, triangles with three equal sides, no two sides equal, etc. A diagram such as Illus. 1.4 illustrating relationships among sets is called a Venn diagram.

Many sets are defined by a property which all the elements in the set must possess. It is possible, though, that no elements have the defining property. The set containing all cars manufactured by the Ford Motor Company before 1875 is an example. This set contains no elements; that is, no cars were manufactured by Ford before 1875.

Definition 1.4   The set which contains no elements is called the *empty* or *null set* and is denoted by the symbol ϕ.

Empty braces { } are also correct as a means of *listing* no elements. We shall use ϕ in this text.

There is just one empty set and *the empty set is a subset of every set.*

In Illus. 1.5 the two line segments indicated have no points in common; that is, the set of common points contains no elements, or is the empty set.

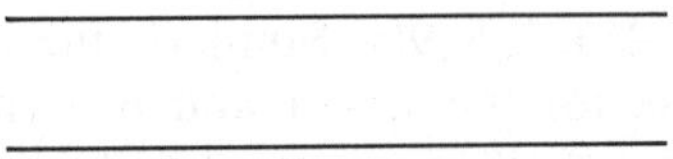

ILLUSTRATION 1.5   Two segments that have no points in common

EXPERIMENT 1B

1   Consider the set of animals as a universal set and the sets of horses, cows, dogs, and cats as respective subsets of the universe. Construct a Venn diagram depicting the universal set and the subsets. Do any of your subsets overlap?

2   Draw two different diagrams interpreting a universe of animals where {cows} and {dogs} are subsets. Use the empty-set concept for one interpretation.

3   Construct a Venn diagram in which $U = \{x \mid x$ is a person} and some subsets of U are students, males, college students, and people with IQ's over 25. Show the proper relations among subsets.

4   Draw a Venn diagram depicting how new model cars can be purchased with one or more of the following options: (1) power steering, (2) air conditioning, (3) stereo tape player.

5   Consider a universe of circles. Let one subset be circles with 6-inch radius, and another subset circles with $\frac{1}{2}$-foot radius. What do you note about these two subsets?

6   Draw a Venn diagram to depict each of Exercises 1A(2) to 1A(11). Do your diagrams help you see when you can draw a conclusion?

EXERCISE 1B

1   If $B = \{2, 4, 6\}$, is B a subset of B according to the definition?

2   For any set D is $D \subseteq D$? Is $D \subset D$?

3   Which of the following sets are empty?
    (a) The set of numbers used in counting
    (b) The set of fingers on your right hand
    (c) The set of triangles which have two sides
    (d) The set of even numbers which may be substituted for x so that $x + 1 = 4$ is true

4   List the elements in each of the following sets.
    (a) The set of positive integers less than 10
    (b) The set of prime numbers less than 20
    (c) The set of letters on a typewriter keyboard needed to spell *Mississippi*
    (d) The set of letters in a Scrabble game needed to spell *Mississippi*
    (e) The set of courses you are enrolled in this quarter

5   Why should your answers to (c) and (d) in Exercise 1B(4) be different?

6   Give reasons for your answers to the following:
    (a) Is $0 \subseteq \phi$?
    (b) Is $\phi \subseteq \phi$?
    (c) If Q is a set, is ϕ a subset of Q?
    (d) Is $9 \in \{8, 9, 10\}$?
    (e) Is $5 \subseteq \{3, 4, 5\}$?

7   List all the subsets of
    (a) $K = \{\$, \#\}$
    (b) $P = \{$line, plane, space$\}$

8   For each of the following find a universal set which contains the given sets as subsets. First use a roster or listing method, then try to find a description.
    (a) {circle, ellipse}, {square, rectangle, parallelogram}
    (b) $\{2, 4, 6\}, \{1, 3, 5, 9\}, \{2, 5, 7\}$
    (c) {red, blue, green}, {yellow, white}
    (d) $\{a, e, i, o, u\}, \{x, y, z\}, \{m, n, o, p\}$

9   Give three descriptions of the null set. Is it true that $\phi = \{\phi\}$?

15

10 For each of the following, give a possible description for the elements in the set.
  (a) {Adam, Eve}
  (b) {quad, quick, quarter, quest}
  (c) {Monday, Tuesday, Wednesday, Thursday, Friday, Saturday, Sunday}
  (d) {Jan. 1, May 30, July 4, Labor Day, Feb. 22, Nov. 11, Thanksgiving Day, Dec. 25}

11 For any two sets A and B is there a difference between knowing that $A \subseteq B$ and knowing that $A \subset B$? Why?

12 If $A \subseteq B$ is it possible for $B \subseteq A$ also?

13 If $A \subset B$ is it possible for $B \subset A$ also?

14 The accompanying diagram shows relationships among several subsets of the set of all angles. Using this diagram, decide if (a) acute angles can be vertical; (b) right angles can be obtuse; (c) right angles can be adjacent.

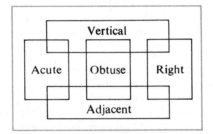

Exercise 1B(14)

15 Use a Venn diagram to show the following: all squares are rhombuses; all rhombuses are parallelograms; and all parallelograms are polygons.

1.6 CLASSIFYING SETS

The two sets $A = \{1, 2, 3\}$ and $B = \{3, 1, 2\}$ contain exactly the same elements. We say these two sets are *equal*.

Definition 1.5 Two sets are *equal* or *identical* if all the elements of one set are elements of the other and vice versa. That is, $A = B$ if and only if $A \subseteq B$ and $B \subseteq A$

The sets $P = \{1, 2, 3\}$ and $Q = \{a, b, c\}$ are not equal even though they have the same number of elements. But it is possible to match each element in P with one element in Q and to match each element in Q with one element in P. Two such sets are said to be in *one-to-one correspondence*.

16

Definition 1.6   Two sets which can be put into one-to-one correspondence are *equivalent* sets.

The two figures in Illus. 1.6 have something in common even though it is not correct to say that they are equal. The commonality derives from

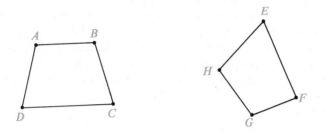

ILLUSTRATION 1.6   A one-to-one correspondence can be established between A, B, C, D and E, F, G, H

the fact that a one-to-one correspondence can be established between the vertices of one figure and those of the other. That is, we can correspond the point A to the point H, correspond the point B to the point G, etc. While the points A, B, C, D are certainly not equal to E, F, G, H, the sets of points are equivalent.

Two equivalent sets are said to have the same *cardinality*; that is, both sets contain the same number of elements. The set $\{a, b, c, d\}$ has four elements. This can be verified by counting as each element is written. The counting would take the form, "One, two, three, four," and the elements in the set would be put into one-to-one correspondence with the first four counting numbers. Thus the numbers of elements in the set would be associated with the number 4. Similarly, the number of elements in the set of your fingers is associated with the counting number 10.

Definition 1.7   If a set can be placed into one-to-one correspondence with the first k counting numbers, then the set is *finite*.

For example, the set $\{a, e, i, o, u\}$ is a finite set because the elements in the set can be corresponded to the numbers 1, 2, 3, 4, and 5. Similarly, the set of people born on January 29, 1950 is a finite set. These people can be given numbers 1, 2, 3,..., k although offhand the value of k is not known.

Sets that cannot be corresponded to the first k counting numbers are called *infinite* sets. The set

$$B = \{1, 2, 3, 4,..., n,...\}$$

17

is an infinite set. It is not possible to count 1, 2, 3,..., up to a fixed number k and in so doing count the very last element in B, no matter how large k is. The set

$$C = \{5, 10, 15,\ldots, 5n,\ldots\}$$

is a proper subset of B. Each element of B can be matched to a single element of C by corresponding the element from B to the element five times as large in C. Similarly, each element of C can be corresponded to an element one-fifth as large in B. This establishes a one-to-one correspondence between the set B and a proper subset of B. This correspondence of a set with a proper subset of itself can only be made when the sets are infinite. Sometimes this property is used as a definition of infinite sets.

EXPERIMENT 1C

1 Explain what set each of the following sets are specifically made to be equivalent to.
(a) A pair of shoes
(b) The fingers on a glove
(c) The spaces in an egg carton
(d) The prongs on the plug on the cord of an electrical appliance

2 Suggest three pairs of equivalent sets from your experiences.

3 We say two sets have the same cardinality, or same number of elements, when they are equivalent. Explain the significance of the following situations in terms of equivalence of sets:
(a) An instructor doesn't call roll when he sees that all assigned seats are filled and no one is standing.
(b) A parking attendant sees his parking lot is full and knows how much money he should have without counting it.
(c) Numbers on a turnstile meter at a public arena and the number of people at a given event in the arena.
(d) A stewardess has a passenger list of 80 names and sees 81 occupied seats on an airline flight.

4 Draw a straight line through a point. Draw a second straight line different from the first through the same point. Draw a third, fourth, and fifth. Is the set of all straight lines drawn through a given point finite or infinite?

5 Mark two points and draw a line segment between them. Can you draw a second line segment, different from the first, between these two points? Is the set of all line segments connecting two points null, finite, or infinite?

6 We have called the set $\{1, 2, 3,\ldots, n,\ldots\}$ the set of counting numbers. Suppose we assign a cardinal number C_0 to this set, and we hear that the cardinality of the integers is also C_0. Suggest how we could verify or disprove this; that is, how can we decide if the number of counting numbers is the same as the number of integers.

18

7 Is the set of people over 20 feet tall empty? Is the set of flying elephants empty? Are these two sets equal? Then, how many different empty sets are there?

EXERCISE 1C

1 In the pairs of sets given below, which are equal?
(a) $A = \{x, y, z\}$; $B = \{X, Y, Z\}$
(b) $A = \{5, 10, 15, \ldots, 5n, \ldots\}$; $B = \{2, 4, 6, \ldots, 2n, \ldots\}$
(c) $A = \{\emptyset\}$; $B = \emptyset$
(d) $A = \{5\}$; $B = \{y \mid 2y - 3 = 10\}$
(e) $A = \{1, 4, 9, 16, \ldots, n^2, \ldots\}$; $B = \{1, 8, 27, \ldots, n^3, \ldots\}$
(f) $A = \{3, 6, 9\}$; $B = \{9, 3, 6\}$
(g) $A = \{@, \#, \%\}$; $B = \{@, \%, \#\}$

2 In Exercise 1C(1), which pairs of sets are (a) equivalent, (b) finite, (c) infinite?

3 Are two equal sets equivalent? Are two equivalent sets equal?

4 (a) Write a set which is equal to {point, line, plane}.
(b) Write a set which is equivalent but not equal to {point, line, plane}.

5 Prove that the set {class, book, pupil, teacher} is not infinite.

6 Decide whether each of the following sets is finite, infinite, or empty.
(a) The set of numbers divisible by 3
(b) The set of prime numbers less than 20
(c) The set of prime numbers between 32 and 36
(d) The set of women presidents of the United States
(e) The set of circles that can be drawn on this page with radius $\frac{1}{4}$ inch and with center on this dot ·
(f) The set of points on this page
(g) The set of all lines through a given point
(h) The set of all lines parallel to a given line
(i) The set of all lines perpendicular to a given line
(j) The set of all right triangles

7 (a) Is the set of all squares equal to the set of all rectangles?
(b) Is the set of all squares equivalent to the set of all rectangles?

1.7 OPERATIONS ON SETS

Operations in arithmetic—addition, subtraction, multiplication, division —are rules by which one or more numbers are associated with a single number. For example, the operation "addition" associates the number 9 with the pair of numbers 5 and 4. In much the same way we may define operations on sets—rules—which associate a set with a pair of given sets. One of these operations, called the *union* of two sets, is denoted by the symbol $\cup$.

19

Definition 1.8 The *union* of two sets A and B is the set which contains all the elements which are in A or in B or in both; that is, $A \cup B = \{x \mid x \in A \text{ or } x \in B\}$.

To find the union of $A = \{1, 3, 5, 7\}$ and $B = \{2, 5, 9\}$, we "join" them to form $A \cup B = \{1, 2, 3, 5, 7, 9\}$. Notice that 5 is listed just once even though it is an element of both A and B.

If we are given $P = \{a, b, c, d, e\}$ and $Q = \{a, e, i, o, u\}$, then $P \cup Q = \{a, b, c, d, e, i, o, u\}$. Remember that order is not important; thus $P \cup Q$ also equals $\{a, e, i, o, u, b, c, d\}$.

The set of all elements common to sets P and Q is the set $\{a, e\}$. This set is called the *intersection* of P and Q. The symbol used to denote the operation of intersection is $\cap$; thus $P \cap Q = \{a, e\}$.

Definition 1.9 The *intersection* of two sets A and B is the set which contains all the elements which are in both A and B; that is, $A \cap B = \{x \mid x \in A \text{ and } x \in B\}$.

The intersection of two sets may be the null set. The sets $R = \{1, 3, 5\}$ and $S = \{2, 4\}$ have no elements in common; hence $R \cap S = \emptyset$. If $R \cap S = \emptyset$ we say that R and S are *disjoint*.

Another useful comparison of sets can be made using the notion of elements that are *not* in a given set.

Definition 1.10 The *complement* of a given set consists of all elements in the universal set not in the given set. The complement of A is denoted by A'; hence, $A' = \{x \mid x \in U \text{ and } x \notin A\}$.

For example, let $S = \{a, e, i, o, u\}$ where the universal set is the alphabet. S' consists of all non-vowels, and hence is the set of consonants:

$$S' = \{x \mid x \text{ is a consonant}\}$$

If we let U be the interior of the triangle in Illus. 1.7 while C is the set of points on or inside the circle, then C' is the set of points indicated by the shaded region.

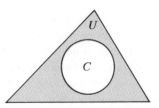

ILLUSTRATION 1.7   The complement of C is the shaded area in U

1.8   THE NUMBER OF ELEMENTS IN A SET

Suppose that set A contains six elements and set B contains five. Is it possible to make any statement about the exact number of elements in $A \cup B$ or in $A \cap B$? Clearly the answer is *no* unless it is also known how many elements sets A and B have in common. There may be as many as 11 elements in the union of the two sets or as few as six. If no element is in both sets, that is if the sets are disjoint, then the number in set A plus the number in set B is the number in the union. Symbolically this is represented as

$$n(A \cup B) = n(A) + n(B)$$

Suppose, however, that set A and set B have two elements the same. Remember the union of the sets does not list an element twice, even though it appears in both sets. In this case the number in the union would be two less than above; nine elements. For two non-disjoint sets A and B the number in the union is given by

$$n(A \cup B) = n(A) + n(B) - n(A \cap B)$$

In like fashion the formula for the number of elements in the union of three sets is found to be

$$n(A \cup B \cup C) = n(A) + n(B) + n(C) - n(A \cap B)$$
$$- n(A \cap C) - n(B \cap C) + n(A \cap B \cap C)$$

If $n(A) = 12$ and $n(B) = 9$ and A and B are disjoint, we can find $n(A \cup B)$. By the first formula above $n(A \cup B) = n(A) + n(B) = 12 + 9 = 21$. If A and B are not disjoint but rather $n(A \cap B) = 5$ then the number of elements in the union is found by using the second formula, that is $n(A \cup B) = n(A) + n(B) - n(A \cap B) = 12 + 9 - 5 = 16$.

EXPERIMENT 1D

1   Assign numbers to the accompanying Venn diagram so that there are 100 in the universe, and (a) no more than 40 in A and no fewer than 60 in B; (b) no more than 80 in A and no more than 80 in B; (c) no fewer than 100 in each of A and B.

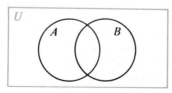

Experiment 1D(1)

2 A waiter at a restaurant notes that over a period of one week he serves 250 dinners. Set *A* represents dinners with appetizer ordered, set *D* represents dinners with dessert ordered, and set *C* represents dinners with cocktails ordered. He noted that during the week he served 75 cocktails, 180 desserts, and 130 appetizers. Also, 27 dinners had all three, 100 had appetizers and dessert, 47 had dessert and cocktails, and 47 had appetizer and cocktails. Explain how this can be possible by using the illustration. How many ordered none of the three extras?

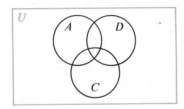

Experiment 1D(2)

3 The illustration depicts the number of elements in the following sets *A* = people in class with blonde hair, *B* = people in class wearing glasses. How many are in each of the following sets?

(a) The Universe
(b) Blondes
(c) People without glasses
(d) People without glasses who are blondes

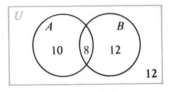

Experiment 1D(3)

4 Draw a Venn diagram to illustrate the following situation. On a class test 23 students made grades of 80 or above, 15 students received grades between 70 and 79, 7 students received grades between 60 and 69, and 7 students made grades below 60. Do you have one circle, two circles, or three circles? Do the circles overlap? If not, could the circles overlap? What is the number of students in the overlapped part?

5 Poll 15 students for their preferences in condiments on hamburgers. Ask them if they prefer mustard, relish, or ketchup, a preference for more than one. Draw a diagram to illustrate the results of your poll.

22

6 Poll 20 students for their preferences in music. Ask them if they prefer classical, jazz, rock, or country and western, limiting the answers to these four categories. Some may indicate a liking for more than one. Draw a diagram to illustrate the results. Is it harder to draw a diagram for four categories than for three?

7 Draw a diagram to illustrate that the subsets books, people, cars, and houses are in a universal set. If these are the only subsets in the universal set, what formula would describe the number of elements in U?

8 Draw a diagram to illustrate the universal set parallelograms, and the subsets squares and rectangles. If R = {rectangles} and S = {squares}, what is the formula for $n(R \cup S)$?

9 Draw a diagram to show that the set of all triangles contains the sets equilateral triangles (three equal sides), isosceles triangles (two sides equal), and scalene triangles (no equal sides) as subsets. If S = {scalene triangles}, E = {equilateral triangles}, and I = [isosceles triangles], what is a formula for $n(S \cup E \cup I)$? For $n(S \cap E \cap I)$?

10 Consider the set of objects in your room. Let A = {objects costing over $10}, B = {clothing items}, and C = {objects that are brown or brown-toned in color}. Do you have any items in each of the following sets?
(a) $A \cup B$ (b) $A \cap B$
(c) $B \cup (A \cap C)$ (d) A'
(e) $(A \cup B)'$ (f) $C' \cap B'$ = $(C \cup B)'$

EXERCISE 1D

1 For each of the following pairs of sets, list the elements in $A \cup B$.
(a) A = {1, 2, 3}; B = {1, 3, 5}
(b) A = {2, 13, 34, 45}; B = {2, 4, 6, 8}
(c) A = {1, 3, 5, 7}; B = {2, 4, 6, 8}
(d) A = $\emptyset$; B = {x, y, z}
(e) A = {1, 2, 3}; B = {x, y, z}
(f) A = {1, 4, 9}; B = {1, 4, 9, 16}

2 In Exercise 1 D(1), find $A \cap B$ for each pair of sets.

3 If set A has four elements and set B has three elements, what is the greatest possible number of elements in $A \cup B$? What do we say of A and B in this case? What is the least possible number of elements in $A \cup B$? What relation is there between A and B when the union is a minimum?

4 If set R has five elements and set S has four elements, what is the greatest possible number of elements in $R \cap S$? What is the least possible number of elements in $R \cap S$? What do we say about R and S when the number of elements in their intersection is zero?

5 Sometimes we can denote the union or intersection of two sets by a single symbol. For example $A \cap U = A$ for all sets A and U. In each of the following give a single word name to the result of the operation.
(a) $A \cup U$ (b) $\emptyset \cap U$

23

(c) $\emptyset \cup U$ (d) $A \cap A$

(e) $A \cup A$ (f) $\emptyset \cap \emptyset$

(g) $A \cup \emptyset$ (h) $U \cap U$

(i) $U \cup U$ (j) $A \cap \emptyset$

6   If $A = \{1, 2, 3, 4\}$, $B = \{4, 5, 6, 7\}$, $C = \{1, 3, 5, 7, 9, 11\}$, and $U = \{1, 2, 3, 4, 5, 6, 7, 8, 9, 10, 11\}$, find

(a) $A \cup B$ (b) $A \cap B$

(c) $A \cup C$ (d) $A \cap C$

(e) $(A \cup B) \cup C$ (f) $A \cup (B \cap C)$

(g) $(A \cap B) \cap C$ (h) $A \cap (B \cup C)$

(i) $A \cup (B \cup C)$ (j) $(A \cap B) \cup C$

(k) $A' \cap U$ (l) $A' \cup U'$

(m) $A' \cup A$ (n) $A' \cap A$

(o) $A' \cap C'$ (p) $(A \cup C)'$

7.   List all possible pairs of sets A and B which satisfy the conditions $A \cap B = \emptyset$ and $A \cup B = \{1, 2, 3\}$ simultaneously.

8   List all possible pairs of sets A and B which satisfy the conditions $A \cap B = \{5\}$ and $A \cup B = \{3, 4, 5\}$ simultaneously.

9   When does $A \cup B = \emptyset$ and $A \cap B = \emptyset$?

10   If $A \cup B = B$, what is the relation between sets A and B?

11   If A is the set of all rectangles, B is the set of all parallelograms, and U is the set of all quadrilaterals, describe

(a) $A \cap B$ (b) $A \cup B$

(c) $A' \cap B$ (d) $A \cap B'$

(e) $(A \cap B)'$ (f) $(A \cup B)'$

12   If A is the set of all right triangles, B the set of all triangles with two equal sides, and U the set of all triangles, describe

(a) $A \cup B$ (b) $A \cap B$

(c) $A' \cup B'$ (d) $A' \cap B'$

(e) $(A \cup B)'$ (f) $(A \cap B)'$

1.9   RELATIONS

In everyday conversation we speak in terms of relations that exist among elements of a given set or between two sets. "Jane *is taller than* Helen" and "Phillip *is in the same class as* John" are two examples of relations among people. More abstractly, these relations can be abbreviated as

$$\text{Jane } (\text{T}) \text{ Helen}$$

and $$\text{Phillip } (\text{S}) \text{ John}$$

where the symbol (T) stands for the relation *is taller than* and the symbol (S) stands for the relation *is in the same class as*. These two statements are exactly alike.in form. Each contains some descriptive relation and a pair of elements upon which the relation is described. Each pair of elements can be considered an *ordered pair*; that is, the elements are given in the order in which they appear in the statement. The ordered pair (Jane, Helen) with the relation Ⓣ yields the statement

Jane *is taller than* Helen

but the ordered pair (Helen, Jane), along with the relation Ⓣ, leads to the statement

Helen *is taller than* Jane.

These two are not equivalent statements. Thus, order is important!

Relations are generally classified according to three properties they either possess or do not possess. If we let Ⓡ represent any relation and let *a*, *b*, and *c* be any elements of a set on which this relation is defined, then we can summarize the three properties as follows:

Property 1.1 (*reflexive property*) If *a* Ⓡ *a*, then the relation is reflexive.

Property 1.2 (*symmetric property*) If *a* Ⓡ *b* implies *b* Ⓡ *a*, then the relation is symmetric.

Property 1.3 (*transitive property*) If *a* Ⓡ *b* and *b* Ⓡ *c* implies *a* Ⓡ *c*, then the relation is transitive.

Writing the properties in this way makes them seem rather abstract; a few examples will help to illustrate them.

Consider the relation *is taller than*, defined on the set of people. To see if this relation is reflexive we ask, "Is it true for any person *p* in the set of all people that *p* Ⓡ *p*, that is, that *p is taller than p?*" Since it is not true that *p* is taller than *p* for any given person *p*, the relation *is taller than* is not reflexive. To test for the symmetric property on the relation *is taller than* we ask, "If *a is taller than b* is it also true that *b is taller than a?*" Since the answer is "no," the relation is not symmetric. The last property, transitivity, is tested by asking, "If *a is taller than b* and *b is taller than c* does this imply *a is taller than c?*" The answer this time is "yes," and we note that the relation *is taller than* has the transitive property.

Definition 1.11 A relation which is not reflexive, is not symmetric, but is transitive is called a *linear order relation*.

Since the relation *is taller than* is not symmetric and is not reflexive but is transitive, it is a linear order relation. As the name implies, any linear order relation puts elements in an order; of course, the type of order depends on the relation.

25

Now consider the relation *is the same weight as* on the set of people. This relation is reflexive. For any given person *a* from the set, it is true that *a is the same weight as a*, or *a* Ⓡ *a*. This relation is also symmetric, for if *a is the same weight as b*, it is also true that *b is the same weight as a*. That is, *a* Ⓡ *b* implies *b* Ⓡ *a*. To test for transitivity we ask, "If *a is the same weight as b* and *b is the same weight as c*, does this imply that *a is the same weight as c*?" The answer is "yes," so this relation has the transitive property.

Definition 1.12 A relation which is reflexive, symmetric, and transitive is called an *equivalence relation*.

Using this definition we can see that the relation *is the same weight as* is an equivalence relation.

EXERCISE 1E

1 Decide in each case if the relation is reflexive, symmetric, and/or transitive.
   (a) *Is married to*, defined on the set of people
   (b) *Is the brother of*, defined on the set of people
   (c) *Is west of*, applied to places in the United States
   (d) *Is not equal to*, defined on the set of positive integers
   (e) *Is richer than*, defined on the set of people
   (f) *Is a sibling of*, defined on the set of people
   (g) *Coincides with*, defined on the set of triangles

2 Name a relation which is
   (a) Reflexive, not symmetric, not transitive
   (b) Reflexive, symmetric, transitive
   (c) Not reflexive, symmetric, not transitive

3 Which of the following are equivalence relations and which are linear order relations?
   (a) *Is smarter than*, defined on the set of people
   (b) *Is the wife of*, defined on the set of people
   (c) *Lives on the same street as*, defined on the set of people
   (d) *Wears the same dress size as*, defined on the set of people
   (e) *Is a cousin to*, defined on the set of people
   (f) *Is less than*, defined on the set of integers
   (g) *Rides on a track 15 feet from the track of*, defined on the set of monorails

4 Can a relation be symmetric and transitive and not be reflexive? Give an example.

5 Given a relation Ⓡ and a set {*a, b, c, d, x*}, such that *a* Ⓡ *b*, *b* Ⓡ *c*, *c* Ⓡ *d*, *d* Ⓡ *x*, and *x* Ⓡ *a*, and such that there are no other pairs that can be related by Ⓡ, decide which of three properties this relation possesses.

26

6   Given a relation Ⓡ on the set {1, 2, 3, 4}, such that 1 Ⓡ 2, 1 Ⓡ 3, 1 Ⓡ 4, 2 Ⓡ 4, 3 Ⓡ 4, and 4 Ⓡ 4, and such that there are no other pairs that can be related by Ⓡ, decide which of the three properties this relation possesses.

7   What kind of relation is suggested by giving dairy products in a grocery store code numbers? By giving them prices?

2 / THE BUILDING BLOCKS OF GEOMETRY

2.1 BASIC IDEAS ABOUT POINTS AND SEGMENTS

In Chapter 1, we saw that we might draw conclusions from experimentation, but if we wish a conjecture to rest on a firm foundation, we must deduce it from what is already known or accepted. Although the geometry in this course will be developed through an experimental process, an attempt will be made to convey some of the feeling of a deductive geometry, particularly at the outset. A deductive geometry starts with undefined terms and assumptions, the foundation on which the rest of the structure is built, and it is with these that our discussion will begin.

In formal geometry, the set of words for which no definition is given includes *point*, *on*, and *between*; there are no simpler concepts available with which to explain them, and it is assumed that the reader has an intuitive understanding of their meanings. For obvious reasons it is desirable that this set of undefined words be as small as possible.

POINT

There are a variety of descriptions that may be given to clarify the notion of a point, such as the sharp end of a pin, the tip of a pencil, or a dot on a sheet of paper. Since a point is only an idea, it is impossible to see one. Euclid's definition of a point translates as, "That which has no parts"; standard geometry texts sometimes say that a point is something that has no width, length, or breadth, or is non-dimensional. Although a point is just an idea, we will characterize a point on the printed page as a dot or as the intersection of two lines.

Using the concept of point, we can make our first geometric definition and start to build on this foundation.

28

Definition 2.1   A *geometric figure* is a non-empty set of points.

Illustration 2.1 represents several geometric figures according to this definition. Notice that no special relation (such as all in a line, grouped

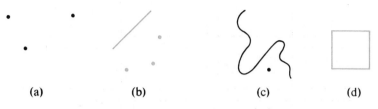

| (a) | (b) | (c) | (d) |

ILLUSTRATION 2.1   Geometric figures

in pairs, etc.) need exist among a set of points to classify it as a geometric figure. This perhaps is a more general idea of a geometric figure than you had previously.

Special figures such as the one shown in Illus. 2.1(d) occupy more of our attention simply because they are more useful. Such a figure may even be common enough to warrant a special name, and so the words square, triangle, rectangle, and circle become part of our language.

Relationships exist between sets of points, and these relationships require mathematical description. But, just as it would be difficult to explain the relation *loves* between a parent and a child, it is difficult to explain many of the relationships among points.

BETWEEN

Certainly you have some notion of what *between* means, but an investigation of this concept shows its complexity. If Mr. Allen, Mr. Baker, and Mr. Chase all own houses on the same side of a street, then you gain a mental picture of the relation of their houses to one another if you are told that Mr. Allen's house is *between* the other two. If the houses are not all on the same side of the street, the situations characterized in Illus. 2.2 can occur. Does your understanding of *between* allow you to say that

| (a) | (b) |

ILLUSTRATION 2.2   Possible positions of three houses on a street

29

Baker's house is not between the other two in these two illustrations? Would you say that Mr. Allen's house is between the other two? If you do think that Allen's house is between the other two, then would you also think that this was true in Illus. 2.3? Actually there is no difference

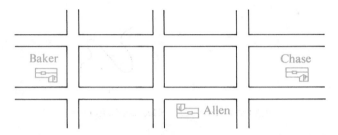

ILLUSTRATION 2.3   A relation of three houses in a city

geometrically between Illus. 2.2(a) and 2.3; Allen's house is not considered between Baker's and Chase's in either case. Geometrically, *between A and B* includes all the unlabelled points in Illus. 2.4(a) but not those shown in

ILLUSTRATION 2.4   (a) Points between *A* and *B*, (b) points not between *A* and *B*

Illus. 2.4(b). What is different about these two pictures? In the first, if we put a straightedge on the page so that its edge was on top of *A* and on top of *B* and use a pencil to draw along the edge from *A* to *B*, the pencil passes over all the unlabelled dots. On the second drawing the pencil would not touch any of the unlabelled dots. Intuitively, the points in Illus. 2.4(a) are between *A* and *B*; those in Illus. 2.4(b) are not. With this understanding we are ready to make another definition.

Without using the term, we have also included another undefined concept, *straight*.

Definition 2.2   A *line segment* is the union of two distinct points *A* and *B* and all points between *A* and *B*.

Illustration 2.5 shows a line segment. Points *A* and *B* have been marked; these two points and all the points between them constitute a line segment. Similarly the points *B* and *C* and all the points between them constitute a line segment. We denote the first of these line segments by $\overline{AB}$ and

30

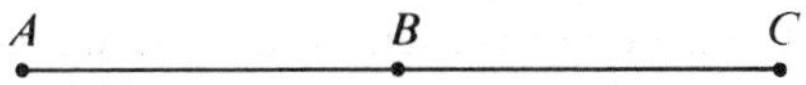

$$A \qquad\qquad B \qquad\qquad C$$

ILLUSTRATION 2.5 $\overline{AB} \cup \overline{BC} = \overline{AC}$; $\overline{AB} \cap \overline{BC} = \{B\}$

the second by $\overline{BC}$. We could also denote them as $\overline{BA}$ and $\overline{CB}$ since order is not important; that is $\overline{AB} = \overline{BA}$ and $\overline{BC} = \overline{CB}$. (Do you see why?) To denote that B is between A and C we write A-B-C. From Illus. 2.5 we can see that $\overline{AB} \cup \overline{BC} = \overline{AC}$ and that $\overline{AB} \cap \overline{BC} = \{B\}$. A and B are called the *endpoints* of the line segment $\overline{AB}$ while all the points between A and B are called *interior* points of the segment. B is an interior point of $\overline{AC}$. The set of points between A and B *without the endpoints* is called an *open segment* and is designated $\overset{\circ\quad\circ}{AB}$. Conversely, a line segment with its endpoints is sometimes called a *closed segment*. A segment denoted by $\overset{\circ\quad}{AB}$ includes endpoint B but not endpoint A, while the segment $\overset{\quad\circ}{AB}$ includes A but not B.

Though we made no definition of point, we must now make several assumptions about points. These assumptions, which we accept without proof, we shall call *postulates*. There will be others throughout the text.

Postulate 2.1   There exist at least two points.

Postulate 2.2   Between any two points there is at least one other point.

These postulates certainly seem reasonable, as we would expect them to be.

With just these two assumptions we can now conclude that there are an infinite number of points. Perhaps it seems strange to state the existence of only three points when intuition suggests that there are many more; but if it is possible to show there is an infinite number of points after postulating just three points, then we need not assume more. Remember, we wish to accept without proof only the bare minimum. Derived conclusions, such as the existence of more points, are called theorems. Thus, we have:

THEOREM 2.1   *There exist an infinite number of points.*

By Postulate 2.1 we know there are two points, say P_1 and P_2. By Postulate 2.2 we know there is a point between them, P_3. Using Postulate 2.2 again, we know there is a point P_4 between P_1 and P_3, a point P_5 between P_1 and P_4, a point P_6 between P_1 and P_5, etc. (see Illus. 2.6).

$$P_1 \; P_6 \; P_5 \qquad P_4 \qquad\qquad P_3 \qquad\qquad\qquad\qquad P_2$$

ILLUSTRATION 2.6   Between two points on a line there exists at least one other point

31

Since this argument can be continued endlessly, we could always find at least one more point between two other points. This suggests that there is an infinite number of points.

An immediate result is another theorem.

THEOREM 2.2 *Between any two points there exists an infinite number of points.*

It is worth mentioning here that although we can derive an infinite set of points by endlessly picking a new point between two others and including this new point in the set we would never completely "fill" any segment of a line. We need to postulate that the line is complete or continuous, that is, that the line has no holes or points missing. More will be said about this assumption later.

In Illus. 2.4(a) the two points A and B had many unlabelled points between them. It takes only a little imagination to see that there are other points which would also lie along the straightedge but are *beyond* either A or B. As a postulate we state

Postulate 2.3 *For any two distinct points A and B there exists at least one point Z such that A-B-Z.*

This postulate leads to the conclusion that there are many points beyond the segment $\overline{AB}$. In Illus. 2.7 $\overline{AB}$ is shown as well as a point Z.

ILLUSTRATION 2.7 The line $\overleftrightarrow{AB}$

Since A and Z are distinct points there is another point Z_1 such that A-Z-Z_1 and another point Z_2 such that A-Z_1-Z_2. Since we can endlessly find another point Z_n such that A-Z_{n-1}-Z_n we conclude that there is an infinite number of points beyond the end of the segment. If we consider all the points Z beyond $\overline{AB}$ we can define a new geometric figure.

Definition 2.3 *A ray consists of a line segment $\overline{AB}$ and all points Z such that A-B-Z.*

In Illus. 2.7, $\overline{AB}$ is a line segment and Z is a point such that A-B-Z. By the definition, the point A and all the points to the right of A constitute a ray. Similarly, A and all points to the left of A constitute a ray. The endpoint and one other point describe a ray; thus ray $\overrightarrow{AB}$ is the ray which has A as an endpoint and includes all points Z such that A-B-Z, as well

ILLUSTRATION 2.8   The union of two rays

as the points on $\overline{AB}$. The arrow is used to differentiate a ray from a segment.

In Illus. 2.8 we have shown the union of the two rays $\overrightarrow{CB}$ and $\overrightarrow{BC}$. This union is a very important new geometric figure.

Definition 2.4   For two distinct points A and B, $\overrightarrow{AB} \cup \overrightarrow{BA}$ is a *line*.

We usually draw a line with arrows at its ends, to indicate that the line extends in both directions. The same convention will be used in specifying lines; thus the line in Illus. 2.7 can be denoted by $\overleftrightarrow{AB}, \overleftrightarrow{AZ}, \overleftrightarrow{XB}$, etc.

Notice that the points A, X, B, and Z are all on $\overleftrightarrow{AB}$. Points located on the same line are said to be *collinear*. The point X on this line can be considered to separate the line into three distinct subsets: the part of the line to the left of X, the point X by itself, and the part of the line to the right of X. We say that the point X separates the line into three disjoint sets of points. All the points on one side of X constitute a *half-line*. Note that there is a small but important difference between a ray and a half-line; the ray includes an endpoint while the half-line does not. We denote a half-line by $\circ\!\!\rightarrow$; hence in Illus. 2.8 the half-line to the right of B is $\overset{\circ}{\overrightarrow{BD}}$.

EXPERIMENT 2A

1   In Illus. 2.8, B and C are two points on $\overleftrightarrow{AD}$. Is it true that $\overline{BC} = \overline{CB}$? On another line locate four points in order from left to right as B, A, D, C. Does the statement $\overline{BC} = \overline{CB}$ still hold? Does it hold if B is between A and D while C is not? Complete the following statement: If B and C are two distinct points on $\overleftrightarrow{AD}$, then _____ = _____.

2   Locate two points A and B and use a pencil to draw $\overline{AB}$. Draw another straight line segment between A and B with a pen. Are these two line segments the same segment? Would you get the same results if you tried this experiment again? Complete the following statement based on your observations: There is (are) _____ line segment(s) between two points. Can you make any statement about the number of lines through two given points?

3   Draw a line. Locate three distinct points on it, labelling them A, B, and C. Is A between B and C? Is B between A and C? Is C between A and B? Did you answer no to two of these questions and yes to one of them? Would your answers always include two no's and one yes no matter how you had labelled the points or where they were located on the line? Complete the

33

following: For three distinct collinear points exactly ____ of them is (are) between the other two.

4 Locate two distinct points A and B. Draw the ray $\overrightarrow{AB}$ and then the ray $\overrightarrow{BA}$. What is the union of these two rays? Would your answer be the same for any two points? Does the order of the points make any difference? Complete this statement: For any two distinct points A and B, $\overrightarrow{AB} \cup \overrightarrow{BA}$ = ____.

5 Locate two distinct points A and B. Draw the ray $\overrightarrow{AB}$ and then the ray $\overrightarrow{BA}$. What is the intersection of these two rays? Would your answer be the same for any two points? Does the order of the points make any difference? Complete the following: For any two distinct points A and B, $\overrightarrow{AB} \cap \overrightarrow{BA}$ = ____.

6 Locate any three non-collinear points. Label them A, B, and C. What is $\overrightarrow{AB} \cup \overrightarrow{AC}$? Is it the same as your answer in Experiment 2A(4)? What is $\overrightarrow{AB} \cap \overrightarrow{AC}$? Is it the same as your answer in Experiment 2A(5)?

7 Locate a point on paper. How many lines can you construct on the paper that will contain that point? Do the same using two points. Three points. What conclusions can you make about the number of points required to determine a specific line.

8 Consider the line segment in the illustration. Is $\overline{AD} = \overline{AD}$? Is $\overline{AD} = \overline{DA}$?

$$A \qquad B \qquad C \qquad D$$

Experiment 2A(8)

9 Draw a line. Choose two points on the line and label them A and B. Now mark a third point C between A and B and a fourth point D between C and B. One way of naming the line with A, B, C, and D on it is $\overleftrightarrow{AB}$. How many other descriptions can you give?

10 Locate three non-collinear points and label them A, B, and C. These points determine three different lines. What are they? Try to arrange three points D, E, and F so just one line is determined. Try to arrange three points G, H, and I so that two and only two lines are determined. What conclusion can you make?

EXERCISE 2A

1 Does the definition for line segment allow a line segment to fit into some other category previously defined? What does your answer suggest about the order of the definitions given in this text and the idea of building new definitions?

2 Draw a line and locate on it, in order from left to right, four points A, B, C, and D; then find
   (a) $\overline{AB} \cup \overline{BC}$ (b) $\overline{AB} \cap \overline{BC}$
   (c) $\overline{AB} \cup \overline{CD}$ (d) $\overline{AB} \cap \overline{CD}$

34

3 Using Illus. 2.8 as a guide, find
 (a) $\overline{BC} \cap \overline{CD}$ (b) $\overrightarrow{BC} \cap \overrightarrow{CD}$
 (c) $\overline{AB} \cap \overline{CD}$ (d) $\overrightarrow{AB} \cap \overrightarrow{CD}$
 (e) $\overrightarrow{BD} \cap \overrightarrow{CA}$

4 In Illus. 2.8 is $\overline{BC}$ a subset of $\overline{AD}$? A proper subset?

5 In Illus. 2.8 is $\overrightarrow{BC}$ a subset of $\overrightarrow{AD}$? A proper subset?

6 Locate on a line, in order from left to right, points M, N, P, and Q; then find
 (a) $\overset{\leftrightarrow}{MN} \cup \overset{\leftrightarrow}{NM}$ (b) $\overset{\circ}{\overrightarrow{PQ}} \cup \overline{PQ}$
 (c) $\overset{\circ}{\overrightarrow{PQ}} \cap \overrightarrow{PN}$ (d) $\overline{MQ} \cup \overset{\circ}{\overrightarrow{PQ}}$

7 When is the intersection of two lines a point? An infinite set? An empty set? Are your answers also true for line segments?

8 (a) Does $\overline{BA} = \overline{AB}$ for any two points A and B?
 (b) Does $\overrightarrow{AB} = \overrightarrow{BA}$ for any two points A and B?
 (c) Does $\overleftrightarrow{AB} = \overleftrightarrow{BA}$?

9 In the diagram what is
 (a) $\overline{AC} \cup \overline{BC}$
 (b) $\overline{BC} \cup \overline{CD}$
 (c) $\overline{AB} \cap \overline{BD}$
 (d) $\overline{AB} \cap \overline{AD}$
 (e) $\overline{BC} \cap \overline{AD}$

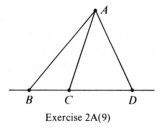

Exercise 2A(9)

10 How many rays on a given line have a particular point A as an endpoint?

11 How many line segments on a line have a particular point A as an endpoint?

12 A carpenter determines a straight line on a wall with a plumb bob. The bob, a pointed metal weight, is attached to a string, and the loose end of the string is tied to a nail high on the wall. This is a physical characterization of a ray where the point where the string is tied to the nail is the endpoint of the ray. Give two other characterizations of rays.

13 Which of the figures shown represent geometric figures?

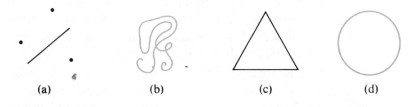

  (a)   (b)   (c)   (d)

35

14 In the illustration name (a) all the line segments, (b) all the rays.

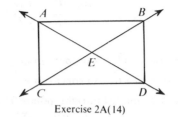

Exercise 2A(14)

15 Draw a line. Locate the five points *A, B, C, D,* and *E* on the line so that all these statements are true at once.

(a) *B* is on $\overrightarrow{AD}$

(b) *A* is on $\overrightarrow{DB}$

(c) *C* is not on $\overrightarrow{BA}$

(d) *E* is on both $\overrightarrow{AB}$ and $\overrightarrow{BA}$

(e) *C* is not on $\overrightarrow{DB}$

2.2 EXPANDING IN DIMENSION

Until now the discussion in this chapter has been confined to one dimension, that is, along a line. However, we live in a three-dimensional world, and we certainly want to include the geometry of more than one dimension in our development. To do this it is necessary once again to make an assumption.

Postulate 2.4   For any line $\overleftrightarrow{AB}$ there is at least one point P that is not on $\overleftrightarrow{AB}$.

This can be symbolized as: there exist *P* and $\overleftrightarrow{AB}$ such that $P \notin \overleftrightarrow{AB}$.

Certainly it is easy to accept this assumption; intuitively we recognize from Illus. 2.1 the existence of points not on a line $\overleftrightarrow{AB}$. Although there may exist many points not on a given line, it is sufficient to postulate the existence of only one. To see how this one point leads us to a new dimension, try the following experiments.

(a) Draw a line segment $\overline{AB}$. Locate some point *P*, not on $\overleftrightarrow{AB}$. Choose a point *X* on $\overleftrightarrow{AB}$ and draw $\overleftrightarrow{XP}$. Choose a second point *Y* on $\overleftrightarrow{AB}$ and draw $\overleftrightarrow{YP}$. Repeat for ten more points on $\overleftrightarrow{AB}$. Describe the resulting figure. Draw several more lines through points on $\overleftrightarrow{AB}$ and *P*. Is it becoming difficult to tell the lines apart? Is it possible to draw still more lines?

(b) Draw the line segment $\overline{AP}$ if you have not done so in (a). Select ten distinct points on $\overleftrightarrow{AP}$ and draw ten lines, one through each of these points and *B*, in a different color. Are any of these new lines the same as those drawn in part (a)? If so, which ones?

36

(c)   Draw the line segment from B to P if you have not done so in (a) or (b). Select five distinct points, W, Q, R, S, and T on $\overrightarrow{PB}$, and draw $\overleftrightarrow{WA}$, $\overleftrightarrow{QA}$, $\overleftrightarrow{RA}$, $\overleftrightarrow{SA}$, and $\overleftrightarrow{TA}$ in a third color.

(d)   On a fresh sheet of paper, repeat the experiment outlined in (a) to (c), this time extending the lines to the edge of the paper. If this process is continued, would all of the paper be covered?

If we could draw every possible line in such an experiment then we would have a two-dimensional figure: a geometric plane. The following definition formalizes the experiments:

Definition 2.5 Given three distinct non-collinear points A, B, and C; locate all lines $\overleftrightarrow{MC}$ where $M \in \overleftrightarrow{AB}$. Locate also the lines $\overleftrightarrow{NA}$ where $N \in \overleftrightarrow{BC}$ and the lines $\overleftrightarrow{QB}$ where $Q \in \overleftrightarrow{AC}$. The geometric plane is $P = \overleftrightarrow{MC} \cup \overleftrightarrow{NA} \cup \overleftrightarrow{QB}$ under the above conditions.

Illustration 2.9 shows how a plane is developed from the definition. Remember that the geometric concept of a plane allows for only two dimensions, while the physicalization of a plane (i.e., a tabletop or a sheet of paper) cannot avoid getting into three dimensions. There is no thickness or depth to a plane; a sheet of paper, no matter how smooth, has a grain, and thus a third dimension to its surface. A plane is not limited in its extents in the two dimensions since it consists of lines of infinite length, but physical objects like the tabletop always have bounds. Hence the concept of "plane" is one we can idealize but never actually encounter, but these differences should not prevent us from using our intuition.

Since a plane figure is a union of lines, it follows that all definitions

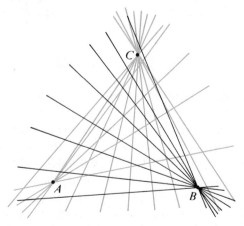

ILLUSTRATION 2.9   Development of a plane

37

concerning line figures hold for plane figures. It does not follow, however, that relationships defined for lines also hold for planes. For instance, on a line one of three distinct points is always between the other two. This is not necessarily true for three points in a plane, as can be seen in Illus. 2.3.

We were able to extend our concept of geometry from a line to a plane by assuming the existence of one additional point. We can extend from two dimensions into three-dimensional space by an analogous assumption:

Postulate 2.5   For a given plane P, there exists a point Q such that $Q \notin P$.

This, along with the two-dimensional properties, is sufficient to yield three-dimensional space. In constructing the second dimension we considered all lines through a point C intersecting a line $\overleftrightarrow{AB}$. Here we shall consider all lines determined by any point $x \in P$ and a point Q not in the plane. Not all such lines can be indicated, but Illus. 2.10(a) shows

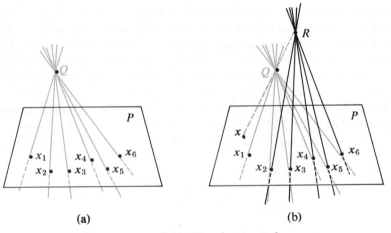

(a)   (b)

ILLUSTRATION 2.10   Development of space

some of the lines through Q and points in the plane. Use your imagination to "fill in" the rest. Drawing all the lines through Q and points in the plane fills almost all of three dimensions or *three-space*. There is only one layer of points, that is, one plane of points, that has not been included. We shall see later that this plane has a special relation to the original plane. If we now pick a second point R, where $R \neq Q$ and $R \in \overleftrightarrow{xQ}$ and repeat the process, drawing all lines through points in the plane and R, we get another family of lines (Illus. 2.10(b)). Though the intersection

38

of these two families of lines is not empty, their union yields three-dimensional space.

Definition 2.6   Given a plane P and a point Q such that $Q \notin P$, consider all points x such that $x \in P$. Let $L = \{m \mid m$ is a line determined by x and $Q\}$. There exists another point R on one m such that x-Q-R. Let $L' = \{n \mid n$ is a line determined by x and $R\}$. Then $L \cup L'$ is three-dimensional space.

We halt here in our intuitive development, but mathematically there is no need to stop. We can easily postulate the existence of a point not in three-space and talk about four-space or the fourth dimension. We *do* run out of dimensions in which to build models and hence to nourish our intuition, but mathematics is not dependent upon models. There are geometries of four and even more dimensions,[1] but we shall not go beyond three dimensions in the remainder of the text.

Just as a point separates a line into three disjoint subsets—two half-lines and the point itself—so a line in a plane separates the plane into three disjoint subsets. The subsets are the two half-planes and the line itself. Illustration 2.11 shows the separation of the plane P by the line $\overleftrightarrow{AB}$. We can carry the analogy to three-dimensional space and consider a plane separating space into three subsets: two half-spaces and the plane.

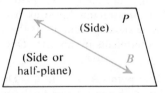

ILLUSTRATION 2.11   A line separates a plane into two half-planes and a line

Many other interesting relationships can be described now that we have extended our discussion to two and three dimensions. The next experiment deals with geometric figures in more than one dimension while developing a familiar concept.

EXPERIMENT 2B

1   Draw a diagram similar to Illus. 2.11 to represent a plane separated into two half-planes and a line $\overleftrightarrow{AB}$. Locate two points C and D in the same half-plane and draw $\overline{CD}$. Is $\overline{CD}$ entirely contained in the same half-plane as C and

[1]H. S. M. Coxeter, *Introduction to Geometry*, John Wiley & Sons, Inc., New York, 1961, Chap. 22.

D? Locate two points Q and R so that Q is in one half-plane and R in the other. Draw $\overleftrightarrow{QR}$. Does $\overleftrightarrow{QR}$ intersect $\overleftrightarrow{AB}$? If two points X and Y are in the same half-plane will $\overleftrightarrow{XY}$ also be in that half-plane? If X and Y are not in the same half-plane will $\overleftrightarrow{XY}$ always intersect the line separating the plane? At how many points will it intersect?

2   Locate three non-collinear points A, B, and C. Will these three points always determine a plane? Is it possible to determine more than one plane with these three points? Illustrate your answer.

3   Choose three non-collinear points A, B, and C. Draw the ray $\overrightarrow{AB}$ and the ray $\overrightarrow{AC}$. What geometric figure is produced? How many points do $\overrightarrow{AB}$ and $\overrightarrow{AC}$ have in common? Repeat for three other points A', B', and C'.

4   In Experiment 2B(3) does the geometric figure produced in each case separate the paper into several disjoint sets? How many? (Remember rays extend without end.) Did you count the points on the rays as one set? Complete these statements: Given three non-collinear points A, B, and C, the rays $\overrightarrow{AB}$ and $\overrightarrow{AC}$ have _____ points in common. Given three non-collinear points A, B, and C, the rays $\overrightarrow{AB}$ and $\overrightarrow{AC}$ separate the plane into _____ disjoint sets.

5   Locate a point B' which is not on the ray $\overrightarrow{AB}$ nor identical with C in the illustration in Experiment 2B(4). Draw $\overrightarrow{AB'}$. Is this geometric figure different from that in Experiment 2B(4)? How does it differ? Would your answer to this question vary depending on where you had chosen B'?

6   Draw $\overleftrightarrow{XY}$. Place a straightedge along $\overleftrightarrow{XY}$. With one end held firmly at X rotate the other end of the straightedge away from you. When you stop, draw a ray along the edge starting from X. Is the figure produced similar to those in the previous experiments in this set? Does the amount of rotation change the figure?

7   Select three non-collinear points A, B, and C. Draw $\overrightarrow{AB}$ and $\overrightarrow{AC}$. Would you describe any points as being *interior* to the resulting geometric figure? Any as being *exterior*? Answer the same questions for three other points A', B', and C'. Would your answers change if the points were collinear?

8   Select three non-collinear points A, B, and C. Draw $\overrightarrow{AB}$ and $\overrightarrow{AC}$. Select some point Q on $\overrightarrow{BC}$ and draw $\overrightarrow{AQ}$. Is the geometric figure determined by $\overrightarrow{AB}$ and $\overrightarrow{AC}$ the same as the one determined by $\overrightarrow{AQ}$ and $\overrightarrow{AC}$? How are they the same? How are they different? If you were to describe a relation between two figures, what would it be? Does your answer depend upon the position of Q?

9   Put one sheet of paper flat on your desk. Place a second piece of paper on top of it so that the two pieces "coincide." Hold one edge of the two sheets together and lift the opposite edge of the top sheet. What would you name the figure formed? Assuming the sheets extend indefinitely, does the figure separate space into disjoint sets? How many?

10   For three non-collinear points A, B, and C draw $\overrightarrow{AB}$ and $\overrightarrow{AC}$. Shade the part of the plane on the side of $\overrightarrow{AB}$ containing C. In another color shade the

40

part of the plane on the side of $\overrightarrow{AC}$ containing B. Shade the intersection of these two sets. How would you relate the final shaded portion to the figure formed by $\overrightarrow{AB}$ and $\overrightarrow{AC}$?

Certainly you have gained some ideas about a geometric figure from the experiments. We shall now give special names to the geometric figures described and formalize the properties illustrated in the experiments.

Illustration 2.12 shows two rays $\overrightarrow{PQ}$ and $\overrightarrow{PX}$ forming an *angle*. In discussing angles we describe them either by naming the common endpoint of the rays or, if necessary for clarity, by naming three points: first, a point on one ray, not its endpoint; second, the common endpoint of the rays; and third, a point on the second ray. The angle in Illus. 2.12 is

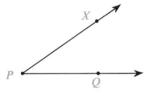

ILLUSTRATION 2.12   Angle XPQ is formed by $\overrightarrow{PQ}$ and $\overrightarrow{PX}$

described as $\angle P$ or as $\angle XPQ$ while the angle in Illus. 2.5 is $\angle ABC$. The rays are called the *sides* of the angle, and the common endpoint is called the *vertex*. When the rays making up the angle are collinear as in Illus. 2.5, the angle is called a *straight* angle. An angle is a set of points, although we intuitively associate the idea of rotation with angles. (Some books even designate angles by $\angle$ to indicate how they are formed.) The concept of angle is easier to define mathematically using sets of points, so the term *angle* here will mean the set of points determined by two rays.

Definition 2.7   Given two rays $\overrightarrow{AB}$ and $\overrightarrow{AC}$ with common endpoint A, the *plane angle* $\angle BAC$ is $\overrightarrow{AB} \cup \overrightarrow{AC}$.

Unfortunately, if we consider an angle as a set of points, the concepts of size and interior and exterior of an angle become more difficult to express. In later chapters we shall fall back on the intuitive concept of rotation to clarify these ideas as needed, that is, when applications to physical objects actually involve motion.

The shaded portion of Illus. 2.13 is the *interior* of the angle; the dotted area is its *exterior*. The words "interior" and "exterior" can only be used in reference to angles that are not straight angles. An angle

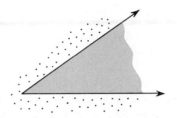

ILLUSTRATION 2.13   An angle divides a plane into three disjoint sets

divides the plane into three disjoint sets: those points interior to the angle, those on the angle, and those exterior to the angle.

We determine the interior of an angle in the following way.

Definition 2.8   Given any non-straight angle *ABC*, the set of points in the plane determined by the intersection of the half-plane on the side of $\overrightarrow{BA}$ containing *C* with the half-plane on the side of $\overrightarrow{BC}$ containing *A* is the *interior of the angle*. (See Illus. 2.14.) The set of points in the plane that is not the interior of the angle or the angle itself is called the *exterior* of the angle.

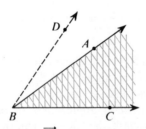

ILLUSTRATION 2.14   Side $\overrightarrow{BA}$ of ∡*ABC* is in the interior of ∡*DBC*

Illustration 2.14 shows ∡*DBA* and ∡*ABC* with *common side* $\overrightarrow{BA}$.

Definition 2.9   Two angles are *adjacent* angles if and only if they have a common side, and the remaining sides (excluding the vertex) are in opposite half-planes with respect to the common side.

By this definition, ∡*DBA* and ∡*ABC* in Illus. 2.14 are adjacent angles. The illustration also displays ∡*DBC* and ∡*ABC*, which have a common vertex and a common side $\overrightarrow{BC}$. But since both remaining sides are in the same half-plane, these angles are not adjacent. We can use ∡*DBC* and ∡*ABC* to define the relation *less than* on the measures of certain pairs of angles.

42

Definition 2.10 If ∢ *ABC* and ∢ *DBC* are such that $\overrightarrow{AB}$ is interior to ∢ *DBC*, then the measure of ∢ *ABC* is less than the measure of ∢ *DBC*. Symbolically, $m(∢ABC) < m(∢DBC)$, where the symbol $<$ represents the relation *less than*.

This definition is limited in scope. Although it is true that $m(∢ ABC)$ is less than $m(∢ DBC)$ in Illus. 2.14, it is possible that a given angle has measure less than another angle when the angles are not positioned as in the illustration. We shall consider such cases later.

A plane angle can also be defined in terms of half-lines. The union of a point *P* and any two half-lines which have *P* as the missing endpoint to both is an angle. This definition agrees with Definition 2.7.

We can extend the discussion of plane angles to angles in three-space, using half-planes. Illustration 2.15 depicts two intersecting half-planes

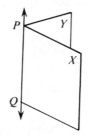

ILLUSTRATION 2.15   A dihedral angle is the union of two half-planes and a line

which have line $\overleftrightarrow{PQ}$ in common. The plane which contains the point *X* can be described as plane *X-PQ* while the plane containing the point *Y* is *Y-PQ*. The union of the two half-planes *X-PQ* and *Y-PQ* and the line $\overleftrightarrow{PQ}$ is the *dihedral angle X-PQ-Y*. Each of the half-planes is called a *face* of the dihedral angle. (Of course, the four points *X*, *Y*, *P*, and *Q* can not all be in the same plane.) The interior of a dihedral angle can be characterized as a wedge cut from a wheel of cheese as shown in Illus. 2.16; the

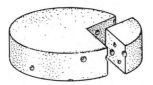

ILLUSTRATION 2.16   A wedge sliced from a wheel of cheese

points in the wedge are interior to the angle, while those in the round are exterior. It must be remembered that, in theory, all points exterior to the angle are included, not just those in the cheese. Again we limit the meaning of interior and exterior angles to angles less than straight angles.

The corner of a room is an example of a dihedral angle, usually a right angle. The wings of an airplane make a dihedral angle with the horizontal when viewed from the front; in fact, this angle is called the *dihedral* of the wings. Illustration 2.17 contains several physical examples of dihedral angles.

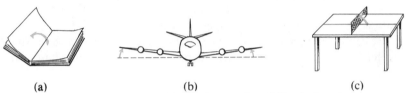

(a) (b) (c)

ILLUSTRATION 2.17 Physical examples of dihedral angles

EXERCISE 2B

1 Find five examples of plane angles in the room you occupy.

2 Find five examples of dihedral angles in the room you occupy.

3 Why is the definition of interior and exterior not applied to straight angles?

4 Is $m(\angle ABC) = m(\angle CBA)$?

5 If $Q \in \overrightarrow{AB}$ and $C \in \overrightarrow{AB}$ is $m(\angle BAC) = m(\angle QAC)$?

6 Define *"less than"* for dihedral angles with a common plane and with terminal planes on the same side of the common plane.

7 Why do you think the word "angle" appears in the phrase, "That guy knows all the *angles*," the word "point" appears in "get the *point*," and "line" is used in "pursuing the same *line* of thought"?

8 In the illustration, name or describe
(a) $\angle ABC \cap \angle BCA$ (b) $\angle DAB \cap \angle EAC$
(c) $\angle DAB \cap \angle BAC$ (d) $\angle ABC \cap \angle DAB$

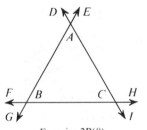

Exercise 2B(8)

9   In the illustration, what points are in the interior of ⦟*AOD*? What segments? What points are in the exterior of ⦟*AOD*? What segments?

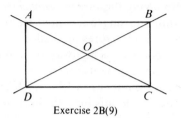

Exercise 2B(9)

10   Consider the three angles ⦟*ABC*, ⦟*DEF*, and ⦟*GHI*.
     (a) Is $m(⦟ABC) < m(⦟ABC)$?
     (b) If $m(⦟ABC) < m(⦟DEF)$, is $m(⦟DEF) < m(⦟ABC)$?
     (c) If $m(⦟ABC) < m(⦟DEF)$ and $m(⦟DEF) < m(⦟GHI)$, does this imply $m(⦟ABC) < m(⦟GHI)$?
     Illustrate your answer. What kind of relation is *less than* on the set of measures of angles?

11   What geometrical advantage is there in building chairs with three legs rather than four?

12   Illustrate all possible results when the following sets of points are intersected:
     (a) A dihedral angle and a plane
     (b) A dihedral angle and a ray
     (c) A half-space and a ray

2.3   POLYGONAL CURVES AND POLYGONS

We turn now to another elementary geometric figure, the polygonal curve. Again, we shall make use of an undefined term and a vocabulary of definitions.

In Illus. 2.18 a set of points called *base points*, indicated by large dots, have been used to construct plane figures consisting of line segments. Each figure consists of more than one segment. In Illus. 2.18(a) the geometric figure consists of just two line segments while the other figures are formed by several connected segments as well as some non-connected

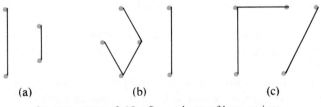

(a)   (b)   (c)

ILLUSTRATION 2.18   Several sets of base points

45

segments. In none of the illustrations have all the base points in the set been joined; that is, each figure consists of at least two parts. We can describe such figures by saying they are *not connected*; we cannot duplicate any of the figures without raising our pencil from the paper some time during the reproduction. Such figures are called *broken line* segments. A dashed line down the center of a highway or a name printed in block letters exemplify broken line segments.

The same sets of base points as in Illus. 2.18 are used to form quite different drawings in Illus. 2.19; this time all the base points in each set are

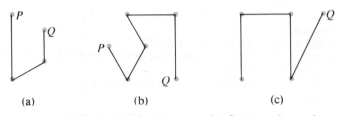

(a)   (b)   (c)

ILLUSTRATION 2.19   Base points connected to form a polygonal curve

joined. It is possible in each case to start at the point *P*, and to trace a path through each of the points in the set to the point *Q* without retracing along a path, without lifting the pencil, and without going through any point more than once. A string of Christmas tree lights is a physical example of a set of points connected by line segments—the light sockets represent points and the wire represents the line segments.

The same sets of base points are shown again in Illus. 2.20, connected this time to result in figures quite different from either previous set. Here

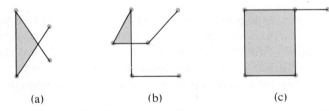

(a)   (b)   (c)

ILLUSTRATION 2.20   Base points connected to form non-simple figures

line segments cross each other or have at least one point in common and, in each of the figures, form an unbroken boundary around some set of points (shaded).

These illustrations indicate several of the ways in which sets of points can be joined; there are, of course, others. But the concept of importance to us now is the difference between the figures in Illus. 2.18 and those in

46

Illus. 2.19 and 2.20. The latter can all be drawn without lifting the pencil from the page, and intuitively we might describe them as being *connected* or *continuous*. The concept of continuity can be defined rigorously, but here we shall use the intuitive concept of connectedness and leave the term undefined. Be sure you understand the difference between Illus. 2.18 and 2.19 or 2.20 before going on.

Definition 2.11　A *polygonal path* (or *polygonal curve*) is the union of a set of one or more line segments such that all are connected end to end in a sequence.

An alternative to this definition is as follows:

Given a set of distinct points $P_1, P_2, \ldots, P_n$, then the union of all line segments $\overline{P_1 P_2}, \overline{P_2 P_3}, \ldots, \overline{P_{n-1} P_n}$, is a polygonal path. Illustration 2.21 displays a polygonal path.

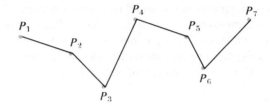

ILLUSTRATION 2.21　A polygonal path

In the figure in Illus. 2.19 no two line segments intersect at a point other than an endpoint; such figures are called *simple*. The figures in Illus. 2.20 are non-simple. We can modify the definition of polygonal path to describe these two difference types of figures.

Definition 2.12　Given a polygonal path consisting of $\overline{P_1 P_2} \cup \overline{P_2 P_3} \cup \overline{P_3 P_4} \cup \cdots \cup \overline{P_{n-1} P_n}$ in a plane (we may also include $\overline{P_n P_1}$), if the intersection of any pair of segments is not empty (exclusive of base points) the path is *non-simple*. Otherwise it is *simple*.

Our set of base points is displayed once again in Illus. 2.22, connected

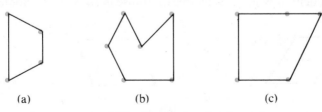

(a)　　　　　　　　(b)　　　　　　　　(c)

ILLUSTRATION 2.22　Simple closed polygonal curves

47

differently. This time a point was selected as starting point, a line segment was drawn from it to another point, a line segment was drawn from there to another point, and so on. The segments were drawn so that no line segment intersected another, and the last segment drawn had the starting point as its endpoint. These are all examples of *closed* figures. Each of these figures is also a simple figure (do you see why?) and consequently they are called *simple closed figures*. We can define a simple closed polygonal curve as follows:

Definition 2.13 Given a set of points P_1, P_2, P_3, ..., P_n where $n \geq 3$, consider the segments $\overline{P_1P_2}$, $\overline{P_2P_3}$, ..., $\overline{P_{n-1}P_n}$, $\overline{P_nP_1}$. If the intersection of the interior points of every pair of segments is empty, then $\overline{P_1P_2} \cup \overline{P_2P_3} \cup \cdots \cup \overline{P_{n-1}P_n} \cup \overline{P_nP_1}$ is a *simple closed polygonal curve*.

Giving such an explicit name to these figures suggests that figures may or may not have the properties of simpleness and closure. A child's hoop, an untwisted rubber band, and a ring are examples of simple closed figures in three dimensions. If a simple closed figure lies entirely in a plane and is the union of line segments, then it is a *simple polygon*. The most familiar polygons are the ones with the fewest sides—triangles and quadrilaterals. Using our concept of sets we may define a triangle as follows:

Definition 2.14 Given any three non-collinear points A, B, and C, then triangle $ABC = \overline{AB} \cup \overline{BC} \cup \overline{CA}$. The points A, B, and C are *vertices*; $\overline{AB}$, $\overline{BC}$, and $\overline{CA}$ are sides.

(This definition of a triangle is the same as the definition of a closed polygonal curve when $n = 3$.) Using the same approach we can define a quadrilateral as follows:

Definition 2.15 Given any four points A, B, C, and D, no three of which are collinear, then a simple quadrilateral is $\overline{AB} \cup \overline{BC} \cup \overline{CD} \cup \overline{DA}$, where the intersection of any pair of open segments is empty. The points A, B, C, and D are the vertices of the quadrilateral, and the four segments are its *sides*.

This definition gives us figures like those in Illus. 2.23. Some familiar

ILLUSTRATION 2.23   Quadrilaterals

48

quadrilateral figures are those we call squares, rectangles, parallelograms, and trapezoids.

EXPERIMENT 2C

1 Consider a set of three non-collinear points. How many ways can they be connected to form different simple polygons?

2 Consider a set of four points, no three of which are collinear. How many simple four-sided polygons can be formed with these points as vertices? Does your answer depend upon the position of the points? Try several arrangements of points.

3 How many other ways are there of joining the four points in Experiment 2C(2) in some order, including closed figures that may cross themselves? Would you consider such a figure a quadrilateral? How could you change the definition of quadrilateral to include such a figure? These figures are called non-simple polygons.

4 Consider five points A, B, C, D, and E. If four of the points are collinear can you draw a simple closed polygon through all the points? If three of the points are collinear can you draw a simple closed polygon through all the points? Does your answer depend upon the position of the other two points? Illustrate with several examples. If at most two points are collinear can you always draw a simple closed polygon?

5 Locate four non-collinear points A, B, C, and D. Draw all line segments connecting pairs of points. Is the resulting figure a simple closed polygon? How many interior regions does this figure have? Compare your answers to those you would give for Illus. 2.22(a). Can you formulate a definition of simple polygons in terms of interior regions?

6 Draw a rectangle. Is this a closed polygonal path? Remove a portion of one side. What kind of path does it now represent? Think of the figure as a wall on a swimming pool; what physical difference would exist in your two drawings as relates to water in the pool?

7 In Experiment 2C(6), if the path represented a fence around a herd of cattle, your first drawing would represent the cattle penned in a closed yard. What physical conditions is represented by your second drawing? Does your answer to this question depend upon the size of the segment removed from one side?

8 In the illustration two points, A and B, have been located. Can you connect them by a polygonal curve that does not intersect the figure? Can you use the answer to this experiment and the last one to formulate another definition for a simple polygon?

Experiment 2C(8)

49

9 Draw a simple polygon. Locate two points inside the polygon. Can you connect these two points with a polygonal curve that does not intersect the polygon? Try this for several different pairs of points. Does your answer depend upon the location of the points or the particular polygon you drew?

EXERCISE 2C

1 Do the points on a line segment fit our concept of connectedness?

2 Consider a rope as a connected set of physical points. If the rope were cut with a very sharp knife, would the rope still all be there? Would it now be connected? What is the difference between a connected set before and after the deletion of just one point.

3 Construct a polygonal curve that is (a) simple but not closed; (b) closed but not simple; (c) neither closed nor simple; (d) simple and closed. What does this mean in terms of the independence of these two properties?

4 Which of the letters of the alphabet represent simple closed curves? Simple polygons?

5 What postulate(s) guarantee(s) a side to a polygon if we define it in terms of points?

6 According to our definitions, which is a rectangle: a tabletop or the boundary of the tabletop?

7 Why is it not possible to have a polygon of fewer than three sides?

8 What is the least number of sides possible for a non-simple polygon in which only two sides intersect?

9 Which of the figures are closed? Simple? Polygons? Polygonal paths?

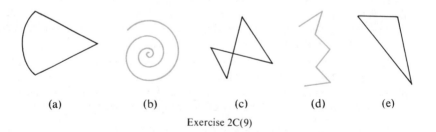

(a)   (b)   (c)   (d)   (e)

Exercise 2C(9)

10 Give two examples of each of the following from your experience.
    (a) Simple closed polygonal path, but not a polygon
    (b) Simple closed polygonal path that is a polygon
    (c) Simple polygon with six sides with one interior
    (d) Broken line path
    (e) Non-closed polygonal path

11 State a definition for a non-simple polygon.

50

12 How would you categorize each of the following in terms of *simple, closed, plane,* or *polygonal* if we extend these concepts to three dimensions?

(a) A bedspring   (b) A telephone pole

(c) A paper clip   (d) A closed safety pin

(e) A fence around a rectangular field

2.4 GENERAL CURVE RELATIONSHIPS

The discussion in this chapter thus far has been limited to geometric figures made up of straight lines. The following definition enables us to broaden our discussion to other types of figures.

Definition 2.16   A *curve* is any connected set of points.

This definition depends on our undefined concept of connectedness or continuity. Informally we have said that any figure that can be drawn without lifting the pencil from the paper is continuous.

Obviously, line segments and polygonal paths are curves, but the definition also includes figures like those in Illus. 2.24. Note that in Illus. 2.24(a) through (c) it is possible to trace the entire curve without

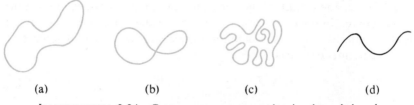

   (a)   (b)   (c)   (d)

ILLUSTRATION 2.24   Curves may or may not be simple and closed

ever lifting the pencil, beginning and ending at the same point. These curves, like some polygons we examined, are *closed* curves, whereas (d) is not closed. In (a), (c), and (d) the curve never crosses itself; these are *simple* curves, whereas (b) is a non-simple curve. All simple curves have two properties in common: they are continuous and do not intersect themselves.

In earlier sections we saw how a point separated a line, and a line separated a plane. Simple closed curves also separate the plane into three disjoint sets: points on the curve, points interior to the curve, and points exterior to the curve.

Definition 2.17   Two portions of a plane are said to be *separated* by a curve if and only if there exists a pair of points x_1, x_2, one in each portion, such that every polygonal path connecting x_1 and x_2 intersects the curve.

51

Illustration 2.25 shows the definition being applied to three curves. Obviously, a line separates the plane. In (b) the closed curve separates the portions of the plane containing x_1 and x_2. The path drawn with a dashed line surely intersects the curve, as would any other polygonal path

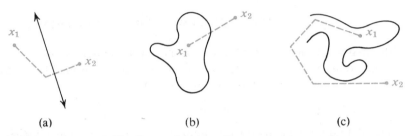

(a)   (b)   (c)

ILLUSTRATION 2.25   Two points in a plane can be separated by a curve

connecting the two points. In (c) we have assumed that x_1 and x_2 are in different portions of the plane, but the polygonal path shown does not intersect the curve; hence the plane is not separated as we supposed. It is not separated at all, for the curve is not closed.

We see, then, that a line separates the plane, as does every simple closed curve. In the latter case, points not on the curve are either exterior to it, or interior. The next definition gives us a way to find *some* of the points exterior to a curve.

Definition 2.18   A point z is *exterior* to a curve C if a ray can be drawn, with z as endpoint, that does not intersect C.

The point z in Illus. 2.26(a) is exterior to C. Now, if we know one exterior point, we can determine whether any other point is exterior or interior (provided it is not on the curve):

Definition 2.19   Consider a point x not on a curve C and a point z exterior to C. If there exists at least one polygonal path from x to z that does not intersect C then x is an *exterior* point of C; otherwise it is an *interior* point.

In Illus. 2.26(b) x is an exterior point because it can be connected to an exterior point z by a polygonal path that doesn't intersect the curve. Point y is interior to curve C. If a man is placed in a maze he can determine whether it is possible to get out (i.e., if he is really on the outside) by walking along, changing directions when necessary (a polygonal path), until he comes to a place where he can proceed in a straight line indef-

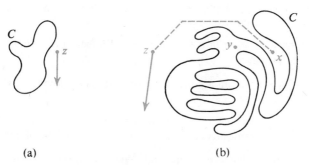

(a)   (b)

ILLUSTRATION 2.26   *x* is an interior point in each curve

initely without bumping into a wall of the maze. A man in a locked room
knows he is separated from the outdoors by the simple observation that
any path he chooses to follow will intersect the boundary.

We can consider three-dimensional separation as an extension of the
concept of separation in two dimensions. How can we separate two birds
which are mobile in three dimensions?

The extension of a line into three dimensions is a plane; thus we can
separate three-dimensional space by using planes or portions of planes.
A birdcage consists of four or more portions of planes and separates
space outside the cage from space inside the cage. Our houses are made
as portions of plane or curved surfaces to separate us from undesirable
phenomena such as rain, excessive heat or cold, bugs, etc. Geometrically
these three-dimensional simple closed figures are called *polyhedra*.

Definition 2.20   Four or more portions of planes determined by polygons
whose pairs of edges are coincident, positioned so that space is separated
into three or more disjoint subsets, constitute a *polyhedron*; each portion
of a plane is termed a *face* of the polyhedron.

We will normally refer to the union of the faces of a polyhedron as
a polygonal surface.

All the concepts of two dimensions discussed here including polygons, simple figures, closed figures, and interior and exterior points have
their three-dimensional counterparts. The following experiments explore
some of these concepts.

EXPERIMENT 2D

1   It has been stated that a point separates a line into three parts. Illustrate this
with a drawing. Is there any other way of separating a line?

53

2 A line separates a plane into two half-planes. Illustrate this. Name at least two other ways a plane can be separated.

3 A plane separates space into two regions. Illustrate this. Find three other ways to separate space.

4 What possible shapes can a polyhedron of four faces assume? What are the shapes of the faces?

5 Is it possible to construct a polyhedron with five faces? Make a diagram. Is it possible to have some faces non-triangular? All faces triangular?

6 Is a non-simple polyhedron (one that separates space into three or more regions) possible? Name some examples.

7 Can a six-sided polyhedron have other than six faces? Other than 12 edges? Other than eight vertices? Consider a cube with one corner sliced off by a plane passing through the midpoints of three segments to that vertex. How many faces, edges, and vertices does this figure have? Now consider a cube with one corner sliced off by a plane passing through two midpoints and another vertex. How many faces, edges, and vertices does this figure have?

EXERCISE 2D

1 Can we define one of the plane regions separated by a line as interior to the line and the other as exterior? Why?

2 Can you phrase in terms of separation the need for the Great Wall of China when it was built?

3 Indicate if each of the following objects separates the plane, assuming that the object is in the plane. Consider only two-dimensional aspects of the object.
    (a) A rubber band   (b) A rope
    (c) A paper clip with ends extended   (d) A spider web
    (e) An open staple   (f) The letters (capitals) of the
                                      alphabet
    (g) A coin   (h) The median strip on a road

4 Indicate which of the following separates space
    (a) A basketball   (b) A carport
    (c) A pyramid   (d) An umbrella
    (e) A piece of paper   (f) A bank vault
    (g) A pencil   (h) The Eiffel Tower

5 Why can't there be fewer than four faces to a polyhedron?

6 Determine which of the figures separates the plane into two regions. Into three regions. Into more than three regions. An arrow is used to indicate that a curve extends without end.

54

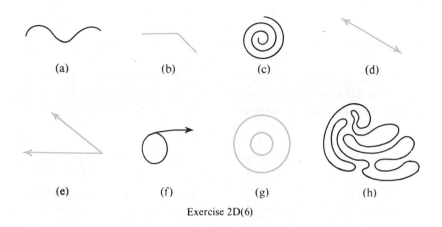

(a)   (b)   (c)   (d)

(e)   (f)   (g)   (h)

Exercise 2D(6)

7 Which of the ten digits 0 through 9 separate the plane into regions? Which of the digits are simple closed curves?

8 What is the name of each of the following polyhedrons?
(a) A child's block   (b) A house with a gable roof
(c) A pyramid   (d) A house with a hip roof

2.5 MORE ABOUT POLYGONS AND POLYHEDRA

There is another property possessed by some simple polygons, the property of convexity. Both figures in Illus. 2.27 are simple quadrilaterals, but Illus. 2.27(b) is not the kind of quadrilateral we usually construct. Such a figure is said to be *concave*, whereas Illus. 2.27(a) is *convex*. The property of convexity or concavity is, of course, not limited to quadrilaterals. Intuitively, adjacent sides of a convex figure form an outward bulge, whereas in a concave figure at least some of the sides seem to collapse inward.

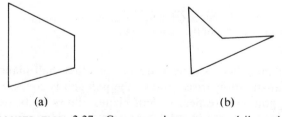

(a)   (b)

ILLUSTRATION 2.27 Convex and concave quadrilaterals

55

Definition 2.21   A simple polygon is *convex* if and only if each of its sides is a segment of a line determining a half-plane which contains the rest of the polygon as a subset. A simple polygon that is not convex is said to be *concave.*

In Illus. 2.28 a polygon is being tested for convexity. In the first sketch side $\overline{AB}$ is extended to form a line in the plane; this line divides the plane into two half-planes such that all the points of the polygon not

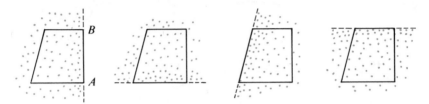

ILLUSTRATION 2.28   Each edge determines a half-plane which contains all the other points of the quadrilateral

on $\overline{AB}$ are in the same half-plane. The other drawings show that the same is true of the other three sides, so this figure is a convex polygon. On the other hand, in Illus. 2.29 a concave polygon is being tested. When side $\overline{PQ}$ is extended to separate the plane, some points of the polygon are

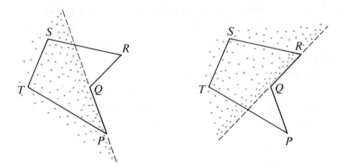

ILLUSTRATION 2.29   The edges $\overline{QP}$ and $\overline{QR}$ determine half-planes which do not contain all the other points of the quadrilateral

in one half-plane whereas others are in the other half-plane If $\overline{QR}$ is tested a similar situation arises, and so the polygon is concave. The location of a polygon with respect to half-planes allows us to define the interior of a convex polygon in a different way.

Definition 2.22 The interior of a polygon is the intersection of all the half-planes that contain some of the segments of the polygon as a subset and have one of the edges of the polygon on the boundary of the half-plane.

Illustration 2.30 shows an interpretation of this definition for a triangle.

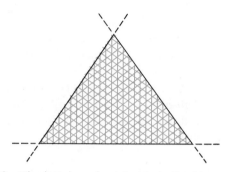

ILLUSTRATION 2.30   The interior of a triangle is the intersection of three half-planes

Polygons can be classified by number of sides. The following names are usually used:

3 sides	triangle
4 sides	quadrilateral
5 sides	pentagon
6 sides	hexagon
7 sides	heptagon
8 sides	octagon
9 sides	nonagon
10 sides	decagon
12 sides	dodecagon

Other classifications of polygons, depending on congruence and/or measure will be discussed in later chapters.

A line segment that joins two vertices of a polygon but that is not an edge of the polygon is a *diagonal*. A diagonal can be defined as a line segment joining two non-adjacent vertices of a polygon. A diagonal need not be interior to the polygon, nor entirely exterior. Illustration 2.31 shows several polygons with diagonals indicated by dashed lines.

Three-dimensional polyhedra have properties of convexity similar to those of two-dimensional polygons. The definition of convexity for a

57

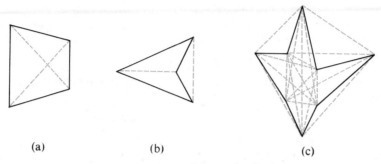

(a)   (b)   (c)

ILLUSTRATION 2.31   Diagonals of polygons

polyhedron is similar to that for a polygon (Definition 2.21), with *half-space* substituted for *half-plane*.   Illustration 2.32(a) shows a convex polyhedron, and (b) a concave polyhedron.   A development of the definitions is left for the exercises.

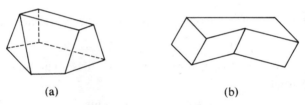

(a)   (b)

ILLUSTRATION 2.32   Convex and concave polyhedra

A diagonal of a polyhedron is a line segment connecting any two of its vertices not in the same face. The diagonals of one convex polyhedron are shown by dashed lines in Illus. 2.33.

ILLUSTRATION 2.33   Diagonals of a polyhedron

EXPERIMENT 2E

1   How many diagonals does a triangle have? Compare this to the number of sides.

58

2 How many diagonals does a convex quadrilateral have? A convex pentagon? A convex hexagon? Compare these to the number of sides in each case.

3 Can you determine a pattern in the answers to Experiments 2E(1) and 2E(2)? Form a generalization about the number of diagonals in a polygon of n sides. Test this generalization for a polygon of eight sides.

4 Do the results of the above experiments change if the polygons are concave rather than convex?

5 Do the results of the above experiments change if non-simple polygons are included as well as simple polygons?

6 Locate seven non-collinear points on a piece of cardboard. Push pins or thumbtacks into the cardboard at these seven points. Loosely loop a piece of string on the cardboard so that all the pins are inside the loop. Tighten the string. The figure produced should be a convex polygon. Do all the pins touch the string, or do some not touch it? Does this give you a new way of deciding if a given set of vertices produces a convex polygon?

7 Given any five points, no three of which are collinear, how many different convex pentagons can you draw with the points as vertices? How many different concave pentagons can you draw using the points as vertices? Try several arrangements of points.

8. In Experiment 2E(7) does your answer change if you locate the points differently?

9 Given any six points, no three of which are collinear, how many different convex hexagons can you draw using the points as vertices? How many concave hexagons? How many non-simple hexagons? Do your answers depend on the positions of the points? Try several arrangements of points.

10 Devise an experiment to determine whether a quadrilateral can have all its sides collapsed. A pentagon? A decagon? Any polygon?

EXERCISE 2E

1 Determine which of the following are concave polygons or polyhedra and which are convex:
  (a) A blackboard   (b) A stop sign
  (c) Your desktop   (d) The base of the Statue of Liberty
  (e) Your home   (f) The boundaries of Colorado
  (g) A baseball diamond   (h) The boundaries of New Mexico
  (i) The UN Building   (j) Your classroom

2 How many diagonals does a tetrahedron have? A pentahedron? A hexahedron? An octahedron? Is there more than one answer to some of these?

59

3 Suppose we define a convex polygon as one in which every two points belonging to it or its interior can be connected by a line segment which also belongs to the polygon and its interior. Compare this definition to the one given in the text. Do you think this is better? Why?

4 Construct a definition for a convex polyhedron.

5 Using the definition you formed in Exercise 2E(4), how would you define a concave polyhedron.

6 Name the following polygons according to their number of sides. Determine the number of diagonals in each.

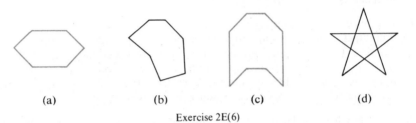

(a)   (b)   (c)   (d)

Exercise 2E(6)

7 Define the interior of a convex polyhedron using a definition analogous to that in Exercise 2E(5).

8 Can a triangle be concave?

9 Can a concave polygon ever have all its diagonals completely exterior to itself? Illustrate if possible.

10 Can a concave polygon ever have all its diagonals entirely interior to itself?

11 How is it possible to arrange ten men in five rows so that each row contains four men? What figure does this arrangement suggest?

12 What kind of a polyhedron is the typical diamond in an engagement ring? Does the answer to this question vary depending on the size of the diamond?

13 Each of the figures shown is a simple closed curve with one point labelled A. Mark another point B on each curve so that $\overline{AB}$ is not entirely interior to the curve. State a general definition for a simple closed curve to be called *concave*.

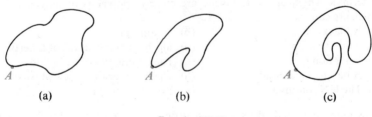

(a)   (b)   (c)

Exercise 2E(13)

14 Using the observations you made in Exercises 2E(13), define a convex curve.

60

3 / SOME RELATIONS IN GEOMETRY

3.1 CONGRUENCE OF GEOMETRIC FIGURES

Of the many relationships among geometric figures, that of congruence probably is of use in more areas of our civilized society than any other. A dressmaker pins a pattern on top of a piece of material and cuts the fabric along the edge of the pattern, in order to make the piece of cut material very close to the pattern in size and shape. If the two pieces are exactly the same size and shape, then the cut material and the pattern are *congruent*. Congruence is basic to any mass production process. An injection-molding machine produces wastebaskets which are all the same in size and shape; they are congruent three-dimensional figures. Every 1″–#4 machine screw must be congruent to every other or replacement is impossible.

When we speak of two figures having the relationship of congruence we mean they are exactly the same size and shape. This is an intuitive, not a formal, definition. Mathematically, congruence is an undefined term. Each pair of figures in Illus. 3.1 is the same size and shape. These sets of figures are *congruent pairs*.

In practical situations minute differences in sizes of similar parts may not affect the functioning of the parts, and we can consider them to be congruent. However, the mathematical *idea* of congruence requires that there not be even the slightest variation in the dimensions of congruent figures. We could measure the two segments in Illus. 3.1(a), and if their lengths were exactly the same, we could say they were congruent. It would be more difficult to determine if the figures in Illus. 3.1(b) were congruent by measuring. As an alternative we might say that the two figures are

61

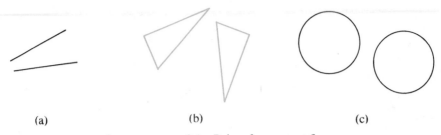

(a)   (b)   (c)

ILLUSTRATION 3.1   Pairs of congruent figures

congruent if they can be made to coincide. This might give us a practical means for determining congruence of physical objects, but as yet we have no way of moving a geometric figure. Whatever means we use to show congruence, the point is that the correspondence or "sameness" must be exact and complete.

Since line segments have one dimension only, congruence of segments must relate to size. Similarly, angular congruence relates to shape. Other congruences are comprised of combinations of these. The next set of exercises and experiments should help you gain some insight into this concept. The symbol ≅ will be used to represent the relation "*is congruent to.*"

Two tools used by geometers to explore congruence are the straightedge and the compass. These tools of classical geometry serve as physical devices to carry measurements from one figure to another. We are assuming that as an object is moved in space it does not change shape or size.

We can construct a line segment congruent to a given segment $\overline{AB}$ with a straightedge and compass. (Note that once a compass is set in a fixed position we assume it always produces pairs of points *exactly* the same distance apart.)

Segment $\overline{AB}$ is given. Draw a ray with endpoint C. Fix the compass so that the metal point is on A and the marking point is on B (Illus.

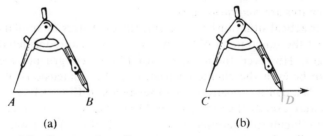

(a)   (b)

ILLUSTRATION 3.2   Constructing a line segment congruent to a given line segment

62

3.2(a)). This represents the measure of segment $\overline{AB}$. Place the point of the compass at C and strike an arc which intersects the ray from C (Illus. 3.2(b)). Label the point D where the arc and ray intersect. We conclude $\overline{CD} \cong \overline{AB}$.

The properties of linear congruence are explored in the following experiments.

EXPERIMENT 3A

1 Draw a line segment $\overline{AB}$. Is $\overline{AB} \cong \overline{AB}$? Would this be true for any segment $\overline{AB}$? Is $\overline{BA} \cong \overline{AB}$?

2 Consider the line segments $\overline{AB}$, $\overline{CD}$, and $\overline{EF}$ in (a). Is $\overline{AB} \cong \overline{CD}$? Is $\overline{CD} \cong \overline{EF}$? Is $\overline{AB} \cong \overline{EF}$? Answer the same questions for the line segments $\overline{AB}$, $\overline{CD}$, and $\overline{EF}$ in (b). What property of relations is implied?

A ———————— B C ———————— D E ———————— F

(a)

A —— B C —— D E —— F

(b)

Experiment 3A(2)

3 Consider line segments $\overline{AC}$ and $\overline{A'C'}$. Is $\overline{AB} \cong \overline{A'B'}$? Is $\overline{BC} \cong \overline{B'C'}$? Is $\overline{AC} \cong \overline{A'C'}$? What property is implied?

A ———— B ———— C A' ———— B' ———— C'

Experiment 3A(3)

4 Consider $\overline{AB}$ and $\overline{CD}$. Are these line segments congruent? Put one end of your compass on A and the other on B. Without changing the opening of the compass, place the point that was on A on C. How would you relate the size of $\overline{AB}$ and $\overline{CD}$?

A ———————— B C ———————— D

Experiment 3A(4)

5 Repeat Experiment 3A(4) for the line segments below, keeping the same order for the segments. How would you relate $\overline{AB}$ and $\overline{CD}$?

A ———————— B C ———————— D

Experiment 3A(5)

63

6   Suppose two houses are one mile apart on a straight road. What property of congruence corresponds to your experience that it is just as far from the first house to the second as it is from the second to the first?

7   Let the figure represent the locations of three houses. Can you draw any conclusions about the relationships of the congruence of $\overline{AB}$, $\overline{BC}$, and $\overline{AC}$?

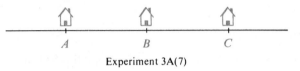

Experiment 3A(7)

8   Consider the same question as Experiment 3A(7) using this figure.

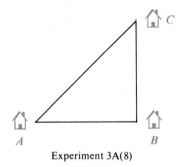

Experiment 3A(8)

9   Given any line segment see if you can devise a method of separating it into two congruent parts using compass and straightedge. This is called *bisecting* the line segment.

10   Given any line segment see if you can devise a method of separating it into three congruent parts (trisecting it) using compass and straightedge.

EXERCISE 3A

The above experiments should help you formulate answers to the first five exercises.

1   Any line segment is congruent to _____.

2   If $\overline{AB} \cong \overline{CD}$ and $\overline{CD} \cong \overline{EF}$ then _____.

3   If *A-B-C* are on a straight line, then $\overline{AB} \cup \overline{BC} \cong$ _____.

4   If A, B, C are any three points not on a line, then $\overline{AB} \cup \overline{BC} \cong$ _____.

5   Given any two segments $\overline{AB}$ and $\overline{CD}$, then either $\overline{AB} \cong \overline{CD}$ or _____.

6   On a ray $\overrightarrow{CP}$ how many points X are there such that $\overline{CX} \cong \overline{AB}$ for some given segment $\overline{AB}$?

64

7 What property of congruence allows world records for the 100-meter dash to be set on different tracks? At different times?

8 What property of congruence gives the furniture mover confidence that a piano 26 inches wide will go through a door he measured to be $26\frac{1}{2}$ inches wide?

9 What property of congruence is a person invoking when he tries to replace something in the position from which he first removed it?

10 Why should the salesman be confident he can return from Chicago to Milwaukee in 2 hours (on a four-lane expressway) if he made the trip from Milwaukee to Chicago in 2 hours?

11–14 Find congruent segments in the following diagrams by using a compass or a pair of dividers

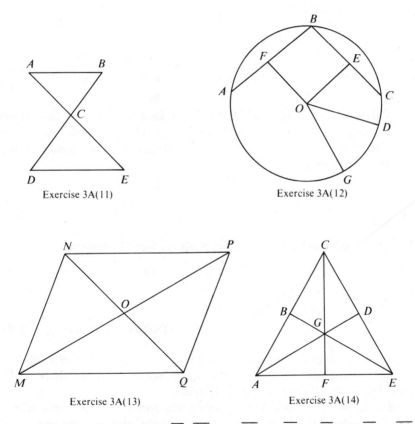

Exercise 3A(11)   Exercise 3A(12)

Exercise 3A(13)   Exercise 3A(14)

15 Consider three line segments $\overline{AB}$, $\overline{CD}$, and $\overline{EF}$. Is $\overline{AB} \cong \overline{AB}$? If $\overline{AB} \cong \overline{CD}$, is $\overline{CD} \cong \overline{AB}$? If $\overline{AB} \cong \overline{CD}$ and $\overline{CD} \cong \overline{EF}$, does this imply $\overline{AB} \cong \overline{EF}$? What kind of relation is $\cong$ on the set of line segments?

65

We hope the experiments and exercises have led you inductively to many conclusions about line-segment congruence.

These properties of congruent segments are stated here as theorems that can be deduced from the assumptions made at the beginning of the chapter and the properties of segments.

THEOREM 3.1 (REFLEXIVE PROPERTY) *For any segment* $\overline{AB}$, $\overline{AB} \cong \overline{AB}$.

THEOREM 3.2 (SYMMETRIC PROPERTY) *For any segments* $\overline{AB}$ *and* $\overline{CD}$, *if* $\overline{AB} \cong \overline{CD}$, *then* $\overline{CD} \cong \overline{AB}$.

THEOREM 3.3 (TRANSITIVE PROPERTY) *For the segments* $\overline{AB}$, $\overline{CD}$ *and* $\overline{EF}$ *if* $\overline{AB} \cong \overline{CD}$ *and* $\overline{CD} \cong \overline{EF}$, *then* $\overline{AB} \cong \overline{EF}$.

THEOREM 3.4 *For a given segment* $\overline{AB}$, *there is exactly one point* X *on the ray* $\overrightarrow{CP}$ *such that* $\overline{AB} \cong \overline{CX}$.

THEOREM 3.5 *For any two segments where A-B-C and A'-B'-C', if* $\overline{AB} \cong \overline{A'B'}$ *and* $\overline{BC} \cong \overline{B'C'}$, *then* $\overline{AC} \cong \overline{A'C'}$.

THEOREM 3.6 *For any two segments* $\overline{AB}$ *and* $\overline{CD}$, *either* $\overline{AB} \cong \overline{CD}$ *or* $\overline{AB} \not\cong \overline{CD}$.

Theorem 3.6 can be expanded if we make a further definition about the relationship of non-coinciding segments. The new definition states precisely what we mean when we say the measure of one segment is less than that of another.

Definition 3.1 The measure of a segment $\overline{AB}$ is *less than* that of a segment $\overline{CD}$ if and only if there exists a point E on $\overline{CD}$ such that C-E-D, for which $\overline{AB} \cong \overline{CE}$.

With this understanding of *less than*, property 6 can be restated as

THEOREM 3.6a (TRICHOTOMY PROPERTY FOR LINE SEGMENTS) *For two segments* $\overline{AB}$ *and* $\overline{CD}$; $\overline{AB} \cong \overline{CD}$, $m(\overline{AB}) < m(\overline{CD})$ *or* $m(\overline{CD}) < m(\overline{AB})$.

3.2 CONGRUENCE OF ANGLES

Besides congruence of size, which is the basis of linear congruence, there is another congruence concept for geometric figures—that of shape. We shall explore this through angle congruence.

In Chapter 2 we saw that an angle was formed by two rays with a common endpoint. Since rays extend without end, we can make no measure of their lengths; therefore "size" has no meaning for angles in terms of lengths. But angles can be different from one another, owing to differences in the amount of "opening" between the rays. Thus, when we say ∡ *A* is congruent to ∡ *B*, we will mean (intuitively) that both angles have the same amount of "opening" between the sides. We can construct an angle congruent to another with straightedge and compass:

To construct an angle congruent to ∡ *B* in Illus. 3.3, draw a ray with endpoint *X*. With one point of the compass at *B*, strike an arc across

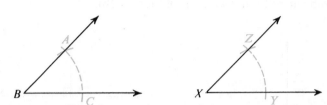

ILLUSTRATION 3.3 Constructing an angle congruent to a given angle

the two rays of ∡ *B* and label the points of intersection *A* and *C*. Without changing the opening of the compass place the point of the compass on *X* and strike an arc which intersects the ray, labeling the point of intersection *Y*. Now with a point of the compass on *C*, set the other point on *A*. Again without changing the opening, set a point of the compass on *Y* and strike an arc that intersects the first arc. Label the point of intersection *Z*, and draw the ray $\overrightarrow{XZ}$. Assuming that the transfer of measures from one figure to the other with the compass is exact, ∡ *ZXY* ≅ ∡ *ABC*.

Note that congruence of angles depends on the relative positions of three points and hence on the congruence of three related line segments. This is where the "shape" concept enters. Whenever we have congruence of *size* and of *shape* between two geometric figures, then we have figure congruence, as suggested earlier.

Being able to construct congruent angles gives us the ability to form right angles, which occur when two lines, or a line and a ray, meet to form two *adjacent* congruent angles. First, we need a new definition for adjacent angles.

Definition 3.2 Two angles are adjacent if and only if they have a common vertex and a common side and the intersection of their interiors is empty.

Compare this definition of adjacent angles with Definition 2.9. (Can you see why this definition was not given in Chapter 2, and how we must build in the definitions?) Illustration 3.4 shows two adjacent angles, ∡ *AMC* and

67

⊀*CMB*, which have a common vertex *M*, a common side $\overrightarrow{MC}$, and no common interior points.

Definition 3.3 If lines $\overleftrightarrow{AB}$ and $\overleftrightarrow{CD}$ intersect at *M* so that ⊀*AMC* ≅ ⊀*CMB*, then the two angles are said to be *right angles*.

It should be clear that there is only one way for two lines to intersect and form right angles. Henceforth, when we wish to designate a right angle it will be marked as in Illus. 3.5(a). Other congruent angles will be designated by identical markings as in Illus. 3.5(b).

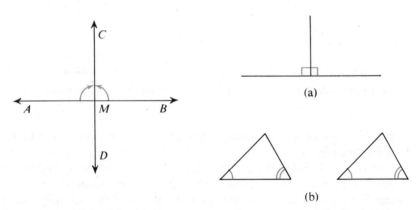

(a)

(b)

ILLUSTRATION 3.4   Lines intersecting to form congruent adjacent angles

ILLUSTRATION 3.5   Designations
for congruent angles

Definition 3.4 Two lines which intersect to form adjacent right angles are *perpendicular*.

A *straight angle* is the angle formed by a straight line, with any interior point of the line considered as the vertex. We see immediately that the non-common sides of two adjacent right angles form a straight angle, with the interior of the angle determined by the context.

EXPERIMENT 3B

1 Construct angles congruent to those shown.

Experiment 3B(1)

68

2 Use a compass to determine if the angles opposite each other formed by a pair of intersecting lines are congruent. Do this for several different figures.

3 Construct figures like those shown. Check with your compass for any congruent angles.

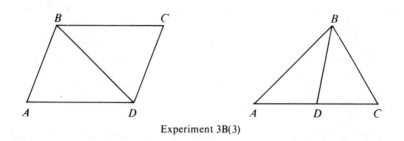

Experiment 3B(3)

4 Which of these pairs of figures would you consider congruent? Did you use methods of determining linear and angular congruence?

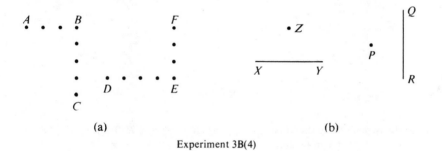

(a)   (b)

Experiment 3B(4)

5 Consider ∡ABC and the ray $\overrightarrow{XY}$ in the illustration. The line $\overleftrightarrow{XY}$ separates the plane into two half-planes, p_1 and p_2. Using $\overleftrightarrow{XY}$ as a base, construct angle $ZXY \cong {∡}ABC$, where $Z \in p_1$. Construct angle $Z'XY$, where $Z' \in p_2$. Are these figures congruent?

Experiment 3B(5)

69

6 Consider the angles *ABC*, *DEF*, and *GHI*. Is ⦤ *ABC* ≅ ⦤*DEF*? Is ⦤*DEF* ≅ ⦤*GHI*? Is ⦤*ABC* ≅ ⦤ *GHI*? What property of the congruence relation does this suggest?

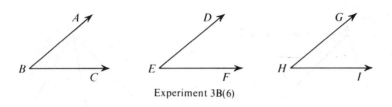

Experiment 3B(6)

7 Consider ⦤*ABC* and ⦤*DEF*. Are these two angles congruent? How would you relate the sizes of two angles? Why?

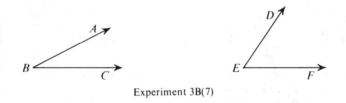

Experiment 3B(7)

8 In the two angles in the illustration, is ⦤*ABD* ≅ ⦤*A'B'D'*? Is ⦤*DBC* ≅ ⦤*D'B'C'*? Can you be sure without checking ⦤*ABC* ≅ ⦤*A'B'C'*?

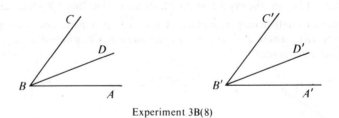

Experiment 3B(8)

9 In triangle *ABC* is $\overline{AB} \cong \overline{AC}$? Is ⦤*ABC* ≅ ⦤*ACB*? Is $\overline{AC} \cong \overline{BC}$? Is ⦤*ABC* ≅ ⦤ *BAC*?

70

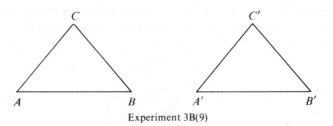

Experiment 3B(9)

10 In triangles ABC and $A'B'C'$ above is $\angle ABC \cong \angle A'B'C'$? Is $\angle BAC \cong \angle B'A'C'$? Is $\angle BCA \cong \angle B'A'C'$?

11 Draw several triangles ABC in which $\overline{AB} \cong \overline{AC}$. Compare the angles opposite $\overline{AB}$ and $\overline{AC}$ in each case. Are these angles congruent?

12 What kind of relation is congruence on the set of angles in terms of properties of relations previously discussed?

13 Given any angle, see if you can devise a method of separating it into two congruent angles using compass and straightedge.

EXERCISE 3B

Use the results of the experiments above to answer the following ten true-false questions.

1 Any angle is congruent to itself.

2 If $\angle A \cong \angle B$, then $\angle B \cong \angle A$.

3 If $\angle A \cong \angle B$ and $\angle B \cong \angle C$, then $\angle A \cong \angle C$.

4 All right angles are congruent.

5 Vertical angles formed by two intersecting lines are not always congruent.

6 The union of any two adjacent angles forms one other angle which uses the exterior sides of the given angles. If we have two more angles congruent to the first two angles, then the union of the second pair of angles will form an angle congruent to the union of the first two.

7 For any pair of angles $\angle A$ and $\angle B$, $\angle A \cong \angle B$.

8 In any triangle with two congruent sides, the angles opposite those sides are congruent.

9 Adjacent angles are congruent angles.

10 Congruent angles are adjacent angles.

11 Show that all four angles formed by perpendicular lines are right angles when you know that two adjacent angles are right angles.

71

12 In the figure, $\overline{OB} \cong \overline{OD} \cong \overline{OC}$; $\measuredangle ABC$, $\measuredangle BCD$, and $\measuredangle BAD$ are right angles. Determine all angles which are congruent to $\measuredangle ODC$. To $\measuredangle OCB$.

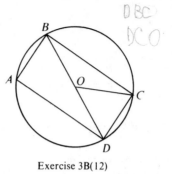

DBC   CBO
DCO

Exercise 3B(12)

13 In the figure, $\overline{AB} \cong \overline{CA}$, $\measuredangle 4 \cong \measuredangle 2$, $\overline{AB} \cong \overline{CD}$, and $\overline{AD} \cong \overline{BC}$. Determine which sets of angles are congruent.

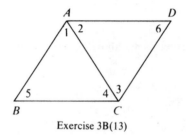

Exercise 3B(13)

It is worthwhile for us to pause here and summarize our ideas about congruence of angles. Perhaps the list below does not include all the conclusions you have reached, but those included are necessary for further discussion.

THEOREM 3.7 (REFLEXIVE PROPERTY) *For any angle A, $\measuredangle A \cong \measuredangle A$.*

THEOREM 3.8 (SYMMETRIC PROPERTY) *For two angles A and B, if $\measuredangle A \cong \measuredangle B$, then $\measuredangle B \cong \measuredangle A$.*

THEOREM 3.9 (TRANSITIVE PROPERTY) *For three angles A, B, and C, if $\measuredangle A \cong \measuredangle B$ and $\measuredangle B \cong \measuredangle C$, then $\measuredangle A \cong \measuredangle C$.*

THEOREM 3.10 *For a given angle ABC there is exactly one half-line $\overset{\circ}{XZ}$ in one half-plane determined by a ray $\overrightarrow{XY}$ such that $\measuredangle ZXY \cong \measuredangle ABC$.*

72

THEOREM 3.11 *If point D is interior to ∡ ABC and point D' is interior to ∡A'B'C', and if ∡ABD ≅ ∡A'B'D' and ∡DBC ≅ ∡D'B'C', then ∡ABC ≅ ∡ A'B'C'.*

THEOREM 3.12 (TRICHOTOMY PROPERTY FOR ANGLES) *For two angles A and B, either ∡ A ≅ ∡ B, m(∡A) is less than m(∡B), or m(∡B) is less than m(∡A).*

Although Chapter 2 included a definition of *less than* for certain angles, the last set of exercises suggests another more general definition for this relation.

Definition 3.5 m(∡ABC) is less than m(∡XYZ) if and only if there exists a point W in the interior of ∡XYZ such that ∡ABC ≅ ∡WYZ.

Angles whose measures are less than that of right angles are called *acute*, while angles whose measures are less than that of straight angles but greater than that of right angles are termed *obtuse*.

3.3 CONGRUENT TRIANGLES

A triangle consists of three line segments situated so that three angles are formed. Since we have discussed congruence of segments and of angles, it is easy to extend our discussion of congruence to triangles. Without going into detail we simply state that two polygons must be exactly alike in size and shape to be congruent; that is, linear and angular congruence must exist for all parts. Congruence of segments in triangles and other geometric figures guarantees the same size, while angular congruence guarantees the same shape.

To denote the triangle with vertices A, B, C, we write $\triangle ABC$. Triangles ABC and $A'B'C'$ are congruent if $\overline{AB} \cong \overline{A'B'}$, $\overline{AC} \cong \overline{A'C'}$, $\overline{BC} \cong \overline{B'C'}$, ∡ $A \cong ∡ A'$, ∡ $B \cong ∡ B'$, and ∡ $C \cong ∡ C'$, and the vertices A, B, C and A',B',C' are labelled in the same order. This means all corresponding sides and corresponding angles of the triangles must be congruent for the triangles to be congruent. The positions of the triangles are not important to congruence; triangles may be in any attitude or even be in a reverse left-to-right position and still be congruent if they meet the criteria. Illustration 3.6 shows three triangles, all congruent to each other.

Although it is essential that all corresponding sides and angles of two triangles be congruent for the triangles to be congruent, it is possible to prove triangle congruence without proving that all corresponding parts are congruent. The next few paragraphs show how to construct triangles when some of the sides and/or some of the angles are given. From such constructions it is easy to determine how many parts, and

73

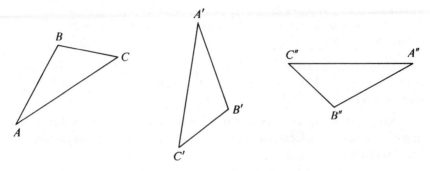

ILLUSTRATION 3.6   Congruent triangles in several positions

which ones, must be known to construct a congruent triangle or to prove congruence.

To construct a triangle given three sides, say *m*, *n*, and *r* in Illus. 3.7(a), draw a ray $\overrightarrow{PQ}$ (Illus. 3.7(b)). Set the compass to the length of *m*, put a point of the compass on *P*, and strike an arc across the ray

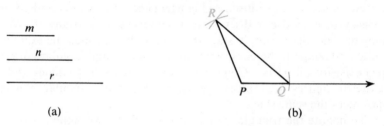

(a)   (b)

ILLUSTRATION 3.7   Constructing a triangle given three sides

at *Q*. Thus $\overline{PQ} \cong m$. Using *P* as center and *n* as length, strike an arc above the ray. Using *Q* as center and *r* as length, strike another arc above the ray. Label the point of intersection of these two arcs *R*. Draw $\overline{PR}$ and $\overline{RQ}$. Line segment $\overline{PR}$ is congruent to *n*, and $\overline{RQ}$ is congruent to *r*; hence $\triangle PQR$ is the desired triangle.

Although the instructions specifically stated that the arcs were to be constructed above the ray, the two arcs could have been constructed below the ray. The resulting triangle would also have sides congruent to *m*, *n*, and *r*.

To construct a triangle given two sides and the angle formed by the two sides, as in Illus. 3.8(a), construct a ray $\overrightarrow{PQ}$ as a base (Illus.

74

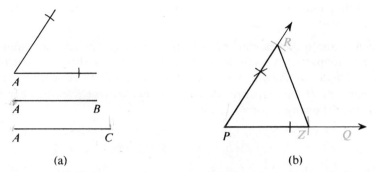

(a)   (b)

ILLUSTRATION 3.8   Constructing a triangle given two sides and the included angle

3.8(b)). On $\overrightarrow{PQ}$ construct an angle congruent to $\angle A$ with its vertex at *P*. Then with *P* as center and with the compass set at length $\overline{AB}$, cut $\overrightarrow{PQ}$ with an arc at *Z*. Then with *P* as a center and with length $\overline{AC}$, cut the terminal side of $\angle P$ with an arc at *R*. Line segment $\overline{ZR}$ completes the triangle.

The illustration shows the construction of the triangle above ray $\overrightarrow{PQ}$, but a similar construction can be made below the ray.

To construct a triangle given two angles and the side between them, as in Illus. 3.9(a), first draw a working line *W* and construct an angle *A'* congruent to $\angle A$ with one side on *W*. Using point *A'* as a center and with the compass set to the length of $\overline{AB}$, mark off segment $\overline{A'B'}$ on *W*. At point *B'* construct an angle congruent to $\angle B$ in the same half-plane as $\angle A$. Extend the terminal sides of the two angles

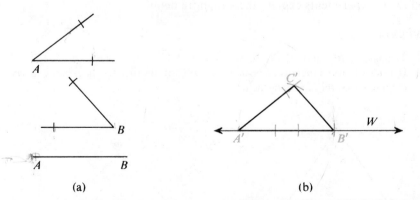

(a)   (b)

ILLUSTRATION 3.9   Constructing a triangle given two angles and the included side

75

until they intersect at C'. This completes the triangle, as shown in Illus. 3.9(b).

Each of the triangles constructed above was completely determined by three of its parts. We should, therefore, be able to show congruence between triangles by showing congruence between only three sets of parts, rather than six. *If the parts are chosen as in the constructions*, this can be done, as the following theorems state:

THEOREM 3.13 (SIDE-SIDE-SIDE THEOREM) *In* $\triangle ABC$ *and* $\triangle A'B'C'$, *if* $\overline{AC} \cong \overline{A'C'}$, $\overline{BC} \cong \overline{B'C'}$, *and* $\overline{AB} \cong \overline{A'B'}$, *then* $\triangle ABC \cong \triangle A'B'C'$.

THEOREM 3.14 (SIDE-ANGLE-SIDE THEOREM) *In* $\triangle ABC$ *and* $\triangle A'B'C'$, *if* $\overline{AB} \cong \overline{A'B'}$, $\overline{AC} \cong \overline{A'C'}$, *and* $\measuredangle A \cong \measuredangle A'$, *then* $\triangle ABC \cong \triangle A'B'C'$.

Although this theorem is stated in terms of $\measuredangle A$ and sides $\overline{AB}$ and $\overline{AC}$, it is not limited to this arrangement, as the naming of the vertices is arbitrary. The important point is that two sides and the angle between them are respectively congruent in the two triangles.

THEOREM 3.15 (ANGLE-SIDE-ANGLE THEOREM) *In* $\triangle ABC$ *and* $\triangle A'B'C'$, *if* $\overline{AC} \cong \overline{A'C'}$, $\measuredangle A \cong \measuredangle A'$, *and* $\measuredangle C \cong \measuredangle C'$, *then* $\triangle ABC \cong \triangle A'B'C'$.

A corollary to Theorem 3.15 that depends on the properties of parallel lines is stated here, though we will develop the necessary justification later.

THEOREM 3.16 (ANGLE-ANGLE-SIDE THEOREM) *In* $\triangle ABC$ *and* $\triangle A'B'C'$, *if* $\measuredangle A \cong \measuredangle A'$, $\measuredangle B \cong \measuredangle B'$, *and* $\overline{AC} \cong \overline{A'C'}$, *then* $\triangle ABC \cong \triangle A'B'C'$.

This says that if any two angles and a side of one triangle are congruent to the corresponding side and angles of a second triangle, then the triangles are congruent. We emphasize again that it is only particular sets of three corresponding parts that determine congruent triangles; the following experiments explore these in more detail.

EXPERIMENT 3C

1 Triangles ABC and $A'B'C'$ have $\overline{AC} \cong \overline{A'C'}$. Is $\triangle ABC \cong \triangle A'B'C'$? If a side of one triangle is congruent to a side of another triangle, can you say definitely that they are congruent?

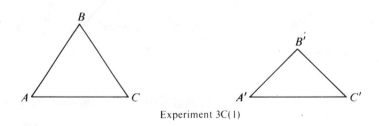

Experiment 3C(1)

2 Draw triangles ABC and $A'B'C'$ with $\overline{AB} \cong \overline{A'B'}$ and $\measuredangle A \cong \measuredangle A'$. Are these triangles necessarily congruent? Draw two triangles which satisfy these conditions that are congruent; draw two that are not congruent.

3 Draw triangles ABC and $A'B'C'$ with $\overline{AB} \cong \overline{A'B'}$, $\measuredangle A \cong \measuredangle A'$, and $\measuredangle B \cong \measuredangle B'$. Are these triangles congruent?

4 Draw triangles ABC and $A'B'C'$ with $\overline{AB} \cong \overline{A'B'}$, $\measuredangle A \cong \measuredangle A'$, and $\measuredangle C \cong \measuredangle C'$. Are these triangles necessarily congruent? Are the triangles congruent if $\overline{AC} \cong \overline{A'C'}$?

5 Construct triangles ABC and $A'B'C'$ with $\measuredangle A \cong \measuredangle A'$, $\measuredangle B \cong \measuredangle B'$, and $\measuredangle C \cong \measuredangle C'$. Are they congruent? Can you construct two triangles which satisfy these conditions and which are not congruent?

6 Construct two triangles which have two angles and a side, not the side between the angles, congruent. Are these triangles congruent? Can you construct two triangles which satisfy these conditions but are not congruent?

7 Some parts of each of the pairs of triangles shown have been marked as congruent. In each case determine from the given information if the triangles are congruent.

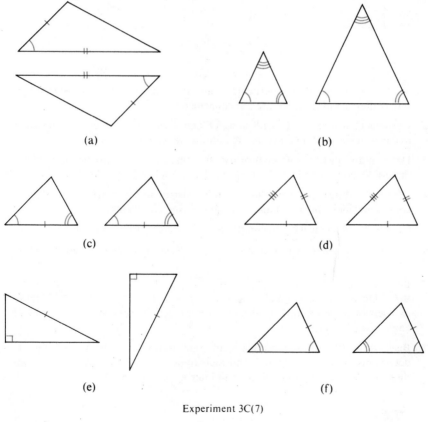

(a)   (b)

(c)   (d)

(e)   (f)

Experiment 3C(7)

77

8 Try to construct a triangle using *m*, *n*, and *r* as sides. Can you make a statement about the relative sizes of the three segments making up a triangle?

Experiment 3C(8)

9 If triangles *ABC* and *A'B'C'* have right angles at *B* and *B'* and $\overline{AB} \cong \overline{A'B'}$, must these triangles be congruent? Construct several sets of triangles which satisfy these conditions. If in addition $\overline{BC} \cong \overline{B'C'}$ are the triangles congruent?

10 In right triangles *ABC* and *A'B'C'* below, two sides of triangle *ABC* are congruent to two sides of triangle *A'B'C'*. Are the two triangles congruent? If not, what further stipulation must be made for the triangles to be congruent?

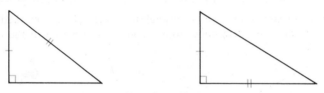

Experiment 3C(10)

11 Construct triangles *ABC* and *A'B'C'* with $\overline{AB} \cong \overline{A'B'}$, $\overline{BC} \cong \overline{B'C'}$, and $\overline{AC} \cong \overline{A'C'}$. Are they congruent? Can you construct two triangles which satisfy these conditions but are not congruent?

12 Construct a triangle *RPQ*, with sides $\overline{RP}$ and $\overline{RQ}$ and ∡ *Q* given. Is it possible to construct more than one triangle fulfilling these conditions?

13 Develop and justify a procedure for the construction of a triangle with all three sides congruent to each other. Such a triangle is said to be *equilateral*.

14 Develop and justify a procedure for the construction of a triangle with two congruent sides. Such a triangle is called an *isosceles* triangle.

15 Construct a triangle with no congruent sides. Such a triangle is called *scalene*.

16 Construct a triangle with one right angle. Such a triangle is called a *right* triangle.

17 Draw triangles *ABC* and *A'B'C'* with $\overline{AB} \cong \overline{A'B'}$, $\overline{BC} \cong \overline{B'C'}$, and $\overline{AC} \cong \overline{A'C'}$. Do you know ∡ *ABC* ≅ ∡ *A'B'C'*? Is ∡ *BCA* ≅ ∡ *B'C'A'*? If these two sets of corresponding angles are congruent, do we also know the third angles are congruent?

18 Bisect the three sides of a triangle to find their midpoints. Draw the segments determined by these midpoints and their opposite vertices, respectively. Are these segments concurrent (do they all intersect at one point)?

78

19 Bisect the three angles of a triangle and extend these rays. Are they concurrent?

The experiments show that if three angles of one triangle are congruent to three angles of another triangle, the two triangles are not necessarily congruent. (There is, however, a definite relationship between the triangles; it will be discussed in a later section on similarity.) They also show that having two sides and a non-included angle congruent in two triangles is not sufficient to determine congruence of the triangles.

EXERCISE 3C

In the first five exercises determine if the given conditions make the triangles congruent. State the theorem named in this section that justifies your conclusion, if it is affirmative.

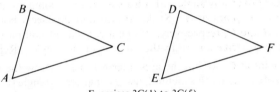

Exercises 3C(1) to 3C(5)

1 $\angle A \cong \angle E, \overline{AC} \cong \overline{EF}, \overline{AB} \cong \overline{DE}$

2 $\angle A \cong \angle E, \overline{DF} \cong \overline{CB}, \overline{DE} \cong \overline{AB}$

3 $\angle C \cong \angle F, \angle B \cong \angle D, \angle A \cong \angle E$

4 $\angle C \cong \angle F, \angle B \cong \angle D, \overline{AC} \cong \overline{EF}$

5 $\overline{AB} \cong \overline{ED}, \overline{DF} \cong \overline{BC}, \overline{AC} \cong \overline{EF}$

6 If two angles and two sides of a triangle are known to be congruent to two corresponding angles and two corresponding sides of another, are the triangles sure to be congruent? Why?

7 If three angles and two sides of a triangle are known to be congruent to three angles and two sides of another not necessarily corresponding, are the triangles sure to be congruent? Why?

8 In Experiment 3A(9) you were asked to bisect a line segment. The following procedure can be used to bisect $\overline{AB}$. Set one point of the compass at A, and strike an arc across $\overline{AB}$ closer to B than A. Repeat, this time starting at B. The arcs intersect at M and N. Construct $\overline{MN}$. This segment bisects $\overline{AB}$.

Justify this procedure using congruent triangle theorems. Is every point on the bisector equidistant from A and B?

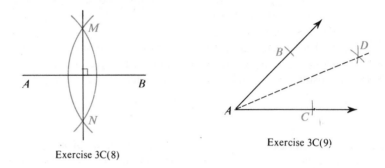

Exercise 3C(8)

Exercise 3C(9)

9   In Experiment 3B(12), you were asked to bisect a given angle. The following is suggested: Given any angle with vertex at A, set a compass at a convenient length with one point at A. Strike an arc that intersects both rays of the angle A at B and C, respectively. Using a convenient compass setting, with one point at B, strike an arc in the interior of the angle. Repeat, with the compass point at C, so that the arcs intersect at D. Draw the ray $\overrightarrow{AD}$. This bisects $\angle BAC$. Justify this method, using congruent triangle theorems.

Find all congruent triangles in each of the following figures. Give justification.

10   O is the midpoint of $\overline{AD}$ and $\overline{BC}$.

11   $\overline{MN} \cong \overline{PQ}$, $\overline{MP} \cong \overline{NQ}$, $\overline{MR} \cong \overline{RQ}$

12   $\angle ABE \cong \angle BED$, $\overline{BD}$ bisects $\angle ABE$, $\overline{AE}$ bisects $\angle BED$.

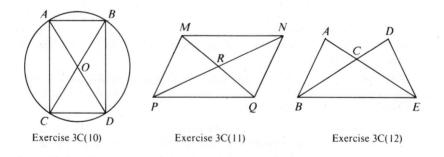

Exercise 3C(10)   Exercise 3C(11)   Exercise 3C(12)

13   If a vertical telephone pole $\overline{AB}$ is held by two equal-length wires $\overline{AC}$ and $\overline{AD}$ which are fastened to a horizontal surface, do we know $\overline{CB} \cong \overline{BD}$? Explain.

80

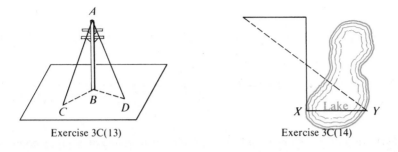

Exercise 3C(13)   Exercise 3C(14)

14 Explain how you could determine the length of $\overline{XY}$ without crossing the lake, using the illustration.

3.4 PARALLELS AND PERPENDICULARS

In the preceding sections we examined congruence of some simple geometric figures. More will be said about this relation later, but it is necessary to develop other relations for lines and planes before we can look at congruence of more complex figures.

The intersection of a pair of lines in a plane may be empty, may contain one point, or may contain an infinite number of points. The last occurs only when the two lines are identical; otherwise two distinct lines intersect in at most one point.

If the two lines intersect in one point, then four angles are formed, and the opposite pairs of angles are congruent. In the special case in which the adjacent angles, and hence all four angles, are congruent, the angles are right angles and the lines are *perpendicular*. We use the symbol ⊥ to indicate perpendicularity.

Using a compass and straightedge we can construct a line perpendicular to a given line at a pre-selected point on the given line, or we can construct a perpendicular to a given line through a point not on the line.

Given a line and some point Z on it. To construct a perpendicular to the line at Z, place one point of a compass on Z and draw two arcs across the line, one on either side of Z, as in Illus. 3.10. Label the intersections of the arcs and the line X and Y. Set the compass to an opening greater than $\overline{XZ}$. Set a point of the compass on X and draw an arc. Without changing the setting of the compass, put a point of the compass on Y and draw an arc that intersects the first. Label this intersection M. Then $\overleftrightarrow{MZ} \perp \overleftrightarrow{XZ}$ at Z.

81

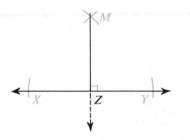

ILLUSTRATION 3.10 Constructing a perpendicular to a line at a given point on the line

We can easily verify that this construction produces a right angle MZY. We know that $\overline{XZ} \cong \overline{ZY}$, $\overline{MZ} \cong \overline{MZ}$, and that $\overline{XM} \cong \overline{MY}$. (Why?) Hence $\triangle XZM \cong \triangle MZY$ and $\measuredangle XZM \cong \measuredangle MZY$. But these two angles together form a straight angle; hence each of them must be a right angle, and $\overleftrightarrow{MZ}$ is perpendicular to $\overleftrightarrow{XY}$.

Given a point P, not on a line L. To construct a perpendicular to L through P, place a point of the compass at P and extend the compass so that it intersects the line L at two points a reasonable distance apart (Illus. 3.11). Label these points M and N. Set a point of the

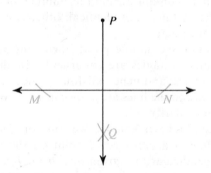

ILLUSTRATION 3.11 Constructing a perpendicular to a line through a point not on the line

compass at M and make an arc in the half-plane determined by L that does not contain P. Without changing the setting of the compass, set its point at N and intersect the arc just made. Label the intersection Q. The line determined by $\overline{PQ}$ is perpendicular to L.

Justification of this construction is left for the exercises. It was mentioned earlier that the intersection of two lines in a plane may be empty.

However, the question of whether or not there exists a line in the plane that does not have points in common with a given line is a philosophical one. Whole geometries have been built on different answers to this question. From a practical viewpoint we can certainly visualize such a line. However, we must accept (or not accept) the existence of such a line without proof. We shall postulate its existence:

Postulate 3.1   Through a point P not on a given line L, there exists one and only one line whose intersection with the given line is empty.

Having assumed the existence of such a line, we can name it.

Definition 3.6   If two lines in a plane have an empty intersection, the lines are parallel.

We use the symbol ∥ to represent this relation between two lines.

Postulate 3.1 characterizes a geometry known as Euclidean geometry, after the Greek mathematician Euclid who, over 2000 years ago, developed many of the properties we are studying here. Though we intuitively think of parallel lines as "lines always the same distance apart" or "lines that never meet," neither of these is a satisfactory definition from a mathematical point of view, since we have not defined distance; nor can we march out along infinite lines to assure ourselves that they never meet.

There are complete and consistent geometries based on other assumptions about parallel lines. These are called non-Euclidean geometries, and some of them have been explored in great detail.

Given any two parallel lines, consider a third line that intersects the other two; Illus. 3.12 shows that two sets of four angles are formed. This situation is common enough so that special names are given to the lines and angles involved. $\overleftrightarrow{EF}$ is called a *transversal* of $\overleftrightarrow{AB}$ and $\overleftrightarrow{CD}$. In general, a transversal is a line which intersects two or more coplanar lines.

Notice that the definition of transversal does not require that the

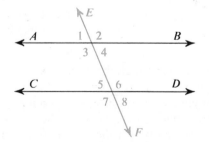

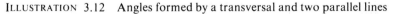

Illustration 3.12   Angles formed by a transversal and two parallel lines

83

lines intersected be parallel; $\overleftrightarrow{EF}$ in Illus. 3.12 is called a transversal in either case. The pairs of angles ∢ 1 and ∢ 5, ∢ 2 and ∢ 6, ∢ 4 and ∢ 8, and ∢ 3 and ∢ 7 occupy corresponding positions and thus are called *corresponding angles*. Angles 3, 4, 5, and 6 are called *interior angles*. Angles 1, 2, 7, and 8 are called *exterior angles*. Thus the pair ∢ 4 and ∢ 5, as well as the pair ∢ 3 and ∢ 6, are *alternate interior* angles; and the pair ∢ 1 and ∢ 7, as well as the pair ∢ 2 and ∢ 8, are *alternate exterior* angles.

One interesting result is a means for determining and constructing parallel lines. For if two lines are parallel the corresponding angles formed by a transversal are congruent. Note that even if one pair of corresponding angles are congruent, then the lines are parallel.

The construction of a line L_2 through a given point P and parallel to a given line L_1 then becomes a matter of copying an angle using point P as vertex. The construction is shown in Illus. 3.13.

Parallel lines may be related by the concept of perpendicularity in the Euclidean geometry we are developing.

THEOREM 3.17 *Given three coplanar lines L_1, L_2, and L_3, such that $L_1 \parallel L_2$, and $L_3 \perp L_1$, then $L_3 \perp L_2$.*

This theorem follows immediately as a consequence of the congruence of corresponding angles. In Illus. 3.14, since $L_3 \perp L_1$, ∢ 1 is a right

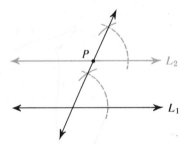

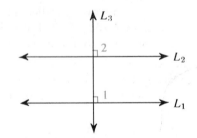

ILLUSTRATION 3.13 Reproducing an angle on a transversal

ILLUSTRATION 3.14 A line perpendicular to parallel lines

angle. Angle 2 is also a right angle because it corresponds to ∢ 1, and $L_1 \parallel L_2$. This in turn implies that $L_2 \perp L_3$.

We have been discussing relationships between two lines in a plane. For two lines in three-space, there is one other possible relationship: their intersection may be empty, yet the lines may not be parallel. Illustration 3.15 shows this possibility. There is a point on each line where a perpendicular may be constructed to both lines, whether or not they are parallel. (Visualize any line in the ceiling of a room, and a line in the floor that passes under the first line.) If two lines in space are not parallel and do not intersect, they are called *skew* lines.

Extending our discussion to three dimensions allows many more possibilities for lines and their relationships to each other and to planes. For example, lines perpendicular to the same line need not be parallel. Illustration 3.16 shows a case in which $\overleftrightarrow{CD}$ and $\overleftrightarrow{EF}$ are both perpendicular to $\overleftrightarrow{AB}$, but $\overleftrightarrow{CD}$ is not parallel to $\overleftrightarrow{EF}$. A few of the more obvious and useful three-dimensional relationships follow.

Definition 3.7 A line is perpendicular to a plane if and only if it is perpendicular to any two lines in the plane through its point of intersection with the plane.

Definition 3.8 A line is parallel to a plane if it does not intersect any line in the plane.

Since any line lies in a plane, Definition 3.8 leads us to the concept of parallel planes. We can consider planes parallel if they have a common perpendicular line (or plane). Illustration 3.17 shows parallel planes X and Y and a line $\overleftrightarrow{AB}$ perpendicular to both of them.

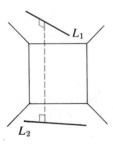

ILLUSTRATION 3.15
Skew lines

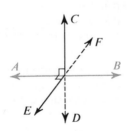

ILLUSTRATION 3.16
Perpendicular lines in
three-space

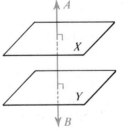

ILLUSTRATION 3.17
A line perpendicular to
parallel planes

Two planes may intersect in a line, have every point in common, or be parallel. Three planes may intersect in a line or a point, have every point in common, or have an empty intersection. Some of these possibilities are shown in Illus. 3.18.

85

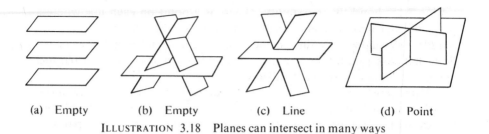

(a) Empty   (b) Empty   (c) Line   (d) Point

ILLUSTRATION 3.18   Planes can intersect in many ways

The intersections of plane surfaces may be polyhedra; the intersections of curved and plane surfaces are more exotic figures such as cylinders and cones. These will be studied separately.

EXPERIMENT 3D

1 Consider a book, a paper, or a desktop as a "plane" and imagine your pencil represents a line.
(a) Hold the "line" so that it does not intersect the "plane."
(b) Hold the "line" so that it intersects the "plane" just once. Can you hold it in several attitudes and achieve this result?
(c) Can you hold the "line" so it intersects the "plane" exactly twice?

2 Draw a line segment L across a sheet of paper. Locate a point p not on L. Make a fold in the paper so that line L is folded back on itself while p is also on the fold. Unfold the paper and draw a line segment m along the fold line. Make a second fold so that m is folded back on itself while p is also on the fold line. Unfold the paper and draw a line segment n along this new fold. Does n appear parallel to L? Why should it be?

3 Suppose you are asked to construct a line from a point P, not on a given line, to a point Q on the line so that the lines will be perpendicular. Try this three times. Can you ever make such a construction? Can you always make this construction?

4 Construct a segment $\overline{AB}$ perpendicular to a segment $\overline{XY}$. Construct a second segment $\overline{CD}$, distinct from $\overline{AB}$, perpendicular to $\overline{XY}$. Identify all the right angles. How many are there? Are some angles formed which are not right angles? How are $\overline{AB}$ and $\overline{CD}$ related?

5 Given a pair of lines, not parallel, draw a transversal and number the angles 1 to 8. Name any pair of angles you think are congruent. Repeat with other pairs of lines. Did you find any congruence between the pairs of angles other than for vertical angles?

6 Draw a triangle. Construct the perpendicular bisector of each side of the triangle. Extend these lines until they intersect. Are they concurrent?

7 Repeat Experiment 3D(6) twice more, using a right triangle and a triangle which has one obtuse angle. What are your results?

8 Construct a square using a method of construcing perpendiculars in this section. How many right angles did you have to construct? See if you can make the square with only one constructed perpendicular.

9 Explain how we know the ironing-board surface in the illustration is parallel to the floor if we know the legs bisect each other. What if the legs are unequal in length but still bisect each other?

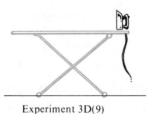

Experiment 3D(9)

10 A builder laying out a foundation for a building assures himself that the foundation has right angles at each corner by finding the midpoint C of one side and noting that $\overline{AC}$ and $\overline{DC}$ are congruent. He already knows that $\overline{AD} \parallel \overline{BE}$ and $\overline{AB} \parallel \overline{DE}$. How does this guarantee right angles?

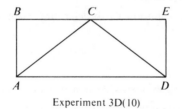

Experiment 3D(10)

11 Construct two congruent equilateral triangles as shown here. Why is $\overline{AB} \parallel \overline{CE}$?

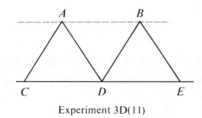

Experiment 3D(11)

87

12 For any triangle *ABC* construct a line through vertex *A* parallel to side $\overline{BC}$. What is the relationship between the angles inside and outside the triangle? Compare the straight angle at *A* to the interior angles. Does this lead you to any conclusion about the interior angles of a triangle?

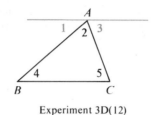

Experiment 3D(12)

13 Repeat Experiment 3D(12) for two other triangles.

14 Consider three parallel lines. What conclusion can you draw about their corresponding angles? Does this give you a way to construct three parallel lines?

15 Suggest a way of determining if two planes (e.g., a floor and a ceiling) are parallel.

16 The union of the interior angles of a triangle is congruent to a straight angle. To what is the union of the interior angles of a quadrilateral congruent? Consider several quadrilaterals.

17 How would you use Definition 3.7 to determine when a pole is perpendicular to a surface?

EXERCISE 3D

1 Using congruent triangles, justify the construction in Illus. 3.11.

2 Show that the alternate interior angles formed by a transversal intersecting two parallel lines are congruent.

3 If two lines in a plane are perpendicular to the same line, are the two lines parallel? Illustrate your answer.

4 If two planes are perpendicular to the same plane, are the two planes parallel? Illustrate your answer.

5 Show that if two lines are parallel or intersect then they must lie in the same plane.

6 Why can't a triangle have parallel sides? Refer to the definition.

7 How many right angles may a triangle have? Why? Refer to Experiments 3D(12) and 3D(13).

88

8 What is the relation between a pair of lines, one contained in each of two parallel planes? Show all possible cases.

9 Experiment 3D(12) suggested that the union of the interior angles of a triangle is congruent to a straight angle. Using the figure, find another way to show this result. $\overline{AB} \parallel \overline{EC}$.

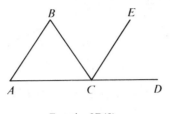

Exercise 3D(9)

10 When are two planes perpendicular?

11 Is it possible for two planes to intersect in a point? Explain.

12 Given three planes P_1, P_2, and P_3, such that $P_1 \cap P_2 = \emptyset$ and $P_2 \cap P_3 = \emptyset$. What are the possibilities for $P_1 \cap P_3$? $P_1 \cup P_3$? $P_1 \cap P_2 \cap P_3$?

13 Justify the angle-angle-side theorem (3.16) using the statement in Exercise 3D(9).

14 Suggest a modification in the side-angle-side theorem (3.14) if one of the angles (not necessarily the included one) is a right angle.

15 In the figure the parallel planes R and S are cut by a third plane P. Name a pair of corresponding dihedral angles. Name a pair of alternate dihedral angles. Name a pair of vertical dihedral angles.

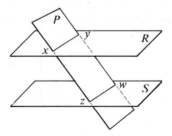

Exercise 3D(15)

16 In the figure $\overline{AC}$ and $\overline{DE}$ are perpendicular to $\overline{AB}$. Which sets of angles are equal?

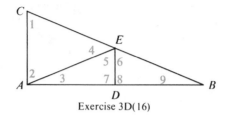

Exercise 3D(16)

3.5 CONGRUENCE OF OTHER FIGURES

It would seem to be a simple matter to extend the concept of triangle congruence to congruence of other polygons by analogy. However, the requirements for congruence become so complex in many-sided polygons that we would need to devote the rest of this book to this topic if we were to investigate it thoroughly. There are, however, several simpler properties which allow insight into the problems of determining congruence in polygons of four or more sides.

Two polygons of *n* sides are congruent if and only if all their corresponding angles and corresponding sides are congruent. For triangles, we need only prove congruence between three special pairs of corresponding parts (out of six) to show that two triangles are congruent, but it is not so simple with quadrilaterals. For example, two triangles are congruent if their corresponding sides are congruent, but two quadrilaterals are not necessarily congruent if their sides are congruent. In Illus. 3.19 the two figures have congruent corresponding sides but are obviously not congruent; their corresponding angles are not congruent. It is sufficient to have the four pairs of sides and one pair of angles congruent to the corresponding parts of another quadrilateral to insure congruency of quadrilaterals, provided the figures are convex (Illus. 3.20).

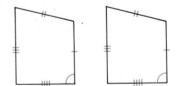

ILLUSTRATION 3.19 Non-congruent quadrilaterals may have corresponding sides congruent

ILLUSTRATION 3.20 Five congruences are needed for two convex quadrilaterals to be congruent

90

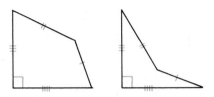

ILLUSTRATION 3.21 Five congruences do not establish congruence in concave quadrilaterals

ILLUSTRATION 3.22 Non-polygonal congruence

However, for concave figures even this is not enough. Illustration 3.21 shows two quadrilaterals with four pairs of congruent sides and one pair of congruent angles, yet these figures are certainly not congruent. The congruence of four-sided figures will be explored further in the experiments. Polygons with more than four sides present even more difficulty.

A different problem arises for figures composed of curves rather than lines (Illus. 3.22). These figures do not fit our definitions about congruence, and we have to go back to our original intuitive notion of congruence as being the quality of having the same size and shape.

EXPERIMENT 3E

1 Draw a convex quadrilateral. See if you can copy this quadrilateral using compass and straightedge by copying one angle and the four sides as suggested in Illus. 3.20. Repeat for concave quadrilaterals.

2 Determine the number of congruent angle pairs necessary to guarantee congruence if only three pairs of sides of two convex quadrilaterals are known to be congruent.

3 Repeat Experiment 3E(2) for the case in which two pairs of sides of the given convex quadrilaterals are known to be congruent.

4 Is it possible to copy a given convex quadrilateral if one side and all four angles are known? Illustrate your answer.

5 Can you suggest a practical method for determining the congruence of two figures such as these?

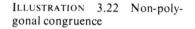

Experiment 3E(5)

6 Construct a four-sided convex polygon with right angles using line segments

91

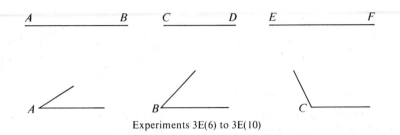

Experiments 3E(6) to 3E(10)

$\overline{AB}$ and $\overline{CD}$ twice each. Is it possible to repeat the experiment and get non-congruent figures?

7 Using $\overline{AB}$, $\overline{CD}$, and $\overline{EF}$ construct a four-sided polygon. Is this the only polygon you can construct with these lines; that is, is the construction unique?

8 Given $\overline{AB}$, $\overline{CD}$, $\overline{EF}$, and $\measuredangle A$ above, construct a four-sided polygon. Is this construction unique?

9 Given $\overline{AB}$, $\overline{CD}$, $\overline{EF}$, $\measuredangle A$, and $\measuredangle B$ above, construct a four-sided polygon. Is this construction unique?

10 Given $\overline{AB}$, $\overline{CD}$, $\overline{EF}$, $\measuredangle A$, $\measuredangle B$, and $\measuredangle C$ above, construct a four-sided polygon. Is this construction unique?

EXERCISE 3E

1 Justify the construction of the congruent quadrilateral in Experiment 3E(1), using congruent triangles.

2 Determine the number of angles needed to assure the congruence of two pentagons if all five pairs of sides are congruent. Justify your answer using congruent triangles.

3 Repeat Exercise 3E(2) for four pairs of congruent sides. Justify.

4 Given two sets of parallel lines as shown, consider the quadrilateral $ABCD$ and the indicated diagonals. Find all pairs of congruent triangles.

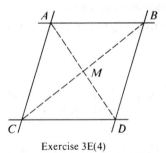

Exercise 3E(4)

5 Using a compass, determine whether or not the figures shown are congruent.

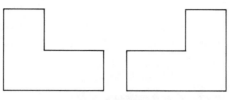

Exercise 3E(5)

6 In each case decide if you would expect the two objects to be congruent.
  (a) A pair of gloves   (b) Two dimes
  (c) Two knives from the same set   (d) Lenses for a pair of glasses
  (e) Two wheels for an automobile   (f) Shirts of the same style and size

7 If your answer to Exercise 3E(5) was "yes" and to Exercise 3E(6a) "no," explain the difference.

8 Using line segments, separate the figure into four congruent regions.

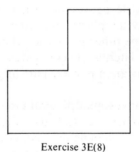

Exercise 3E(8)

9 Give three physical examples of each of the following:
  (a) Two objects which ordinarily would be seen together and are congruent
  (b) Two objects which ordinarily would be seen together and are related but are not congruent

10 How much need you know to determine if two squares are congruent? Two rectangles?

93

4 / GEOMETRIC RELATIONSHIPS APPLIED TO COMMON FIGURES

4.1 PROPERTIES OF QUADRILATERALS

We examined some geometric figures in earlier chapters and classified them according to convexity, simplicity, and other characteristics. In this section we shall discuss some other properties of the more common and useful simple convex quadrilaterals. These figures, with the special names *squares*, *rectangles*, etc., play an important part in geometry and in everyday experiences.

In general, a simple convex quadrilateral (see Illus. 4.1) has noncongruent sides and angles. When we studied triangles, we found that the union of the three angles of a triangle are always congruent to a straight angle. Does a comparable relationship exist for the angles of a quadrilateral? One way to find out would be to construct several different quadrilaterals and compare their angles to see what possibilities exist. How-

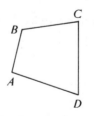

ILLUSTRATION 4.1   A general convex quadrilateral

ever, a simpler method, using a deductive approach, yields a conclusion much more rapidly. For the quadrilateral in Illus. 4.1, the one shown in Illus. 4.2, or, in fact, any convex quadrilateral, we can construct one diagonal such as $\overline{BD}$. This will always produce two triangles, because any two points determine a line segment and the diagonal of a convex quadrilateral is interior to the quadrilateral. In Illus. 4.2 the two triangles are

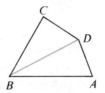

ILLUSTRATION 4.2   A diagonal in a general convex quadrilateral divides it into triangles

$\triangle ABD$ and $\triangle BCD$. Since each of these triangles has interior angles collectively congruent to a straight angle, it follows that the union of the interior angles of any convex quadrilateral is congruent to two straight angles. Thus, we may have a quadrilateral with no, one, two, or four right angles. The figure with four right angles is given the special name *rectangle*. As you read this material you can probably see several examples of rectangular shapes around you.

If a convex quadrilateral has a pair of congruent angles, they may be opposite each other as in Illus. 4.3(a), or they may share a common side as in Illus. 4.3(b). If both pairs of opposite angles are congruent, the figure is

(a)   (b)

ILLUSTRATION 4.3   Two ways a quadrilateral may have a pair of congruent angles

known as a *parallelogram* (see Illus. 4.4(a)). Frequently, parallelograms are defined in terms of their sides, rather than in terms of angles: a parallelogram is a quadrilateral in which both pairs of opposite sides are parallel. In Illus. 4.4(b) parallel lines $\overleftrightarrow{AB}$ and $\overleftrightarrow{CD}$ are cut by parallel lines $\overleftrightarrow{CA}$ and $\overleftrightarrow{DB}$. Angles 1 and 3 are congruent for they are corresponding angles cut off by transversal $\overleftrightarrow{AC}$ on parallel lines $\overleftrightarrow{AB}$ and $\overleftrightarrow{CD}$. Also, ∡1 ≅ ∡4.

95

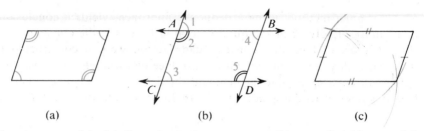

(a)   (b)   (c)

ILLUSTRATION 4.4   (a) Opposite angles congruent, (b) opposite sides parallel, (c) opposite sides congruent

for these are alternate interior angles on the transversal $\overleftrightarrow{AB}$ cutting the parallel lines $\overleftrightarrow{CA}$ and $\overleftrightarrow{DB}$.   Since $\measuredangle 1 \cong \measuredangle 3$ and $\measuredangle 1 \cong \measuredangle 4$, then $\measuredangle 3 \cong \measuredangle 4$; that is, this pair of opposite angles of quadrilateral $ABCD$ are congruent. A similar argument can be used to show $\measuredangle 2 \cong \measuredangle 5$. Hence, if opposite sides of a quadrilateral are parallel, opposite angles are congruent, and vice versa.

Parallelograms can also be defined as quadrilaterals with opposite sides congruent as shown in Illus. 4.4(c). Do you see why? Hence, a quadrilateral is a parallelogram if the opposite pairs of angles are congruent, or if the opposite pairs of sides are congruent, or if the opposite pairs of sides are parallel. Each condition implies the other two. A *rectangle* has opposite sides and opposite angles congruent. Hence every rectangle is a parallelogram, but it is not true that every parallelogram is a rectangle.

It is possible for a quadrilateral to have no congruent sides, two and only two congruent sides, three congruent sides, or all four sides congruent. A pair of congruent sides may be adjacent or opposite, as shown in Illus. 4.5. The figure with two *opposite* sides congruent is the more interesting; if such a figure also has the noncongruent sides parallel as in Illus. 4.6, it is an *isosceles trapezoid*. The segments $\overline{AD}$ and $\overline{BC}$ in the

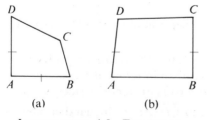

(a)   (b)

ILLUSTRATION 4.5   Two ways a quadrilateral may have a pair of congruent sides

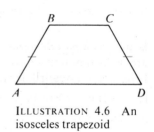

ILLUSTRATION 4.6   An isosceles trapezoid

96

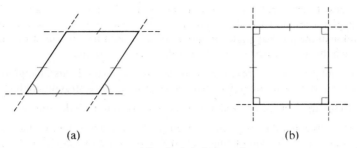

(a) (b)

ILLUSTRATION 4.7   A rhombus and a square

illustration are termed the *bases* of the trapezoid. In the isosceles trapezoid in Illus. 4.6, $\angle BAD \cong \angle CDA$ and $\angle ABC \cong \angle BCD$.

We remove the adjective "isosceles" by removing the congruence condition. That is, in Illus. 4.6, if $\overline{AD} \parallel \overline{BC}$ but $\overline{AB} \not\cong \overline{CD}$, then the figure is still a trapezoid but it is no longer an isosceles trapezoid. Any quadrilateral which has two parallel sides is a *trapezoid*.

If a quadrilateral has two adjacent sides congruent and parallel, it is called a *rhombus* (Illus. 4.7(a)). If we impose one more condition on the rhombus, that is, that one of its angles be a right angle, then it is a *square* (Illus. 4.7(b)). The relationships of the various quadrilaterals to each other and their properties are shown in Illus. 4.8.

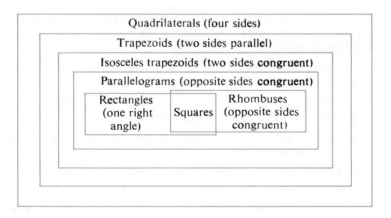

ILLUSTRATION 4.8   Hierarchy of quadrilaterals

EXPERIMENT 4A

1   Given any simple convex pentagon, use the means suggested in this section for determining the number of straight angles that the union of all the interior angles would be congruent to.

97

2 Repeat Experiment 4A(1) for a hexagon, an octagon, a decagon. Can you suggest a method for determining the number of straight angles that the union of the interior angles of any polygon would be congruent to? Did you reach this procedure by inductive or deductive reasoning?

3 If a quadrilateral were allowed to be concave, would this change our conclusion that the union of its angles is congruent to two straight angles?

4 Modify the method devised in Experiment 4A(2) to include concave polygons.

5 Draw two parallel segments. Can you construct a figure that is not a parallelogram by completing a quadrilateral using two congruent segments? Can you do this if the parallel segments are equal?

6 In the figure the two rays are parallel. Construct an angle at *A* using the given ray as one side of the angle and making sure the other side of the angle intersects the ray through *C*. Repeat at *C*, constructing an angle congruent to your first. Is the figure a parallelogram? Why?

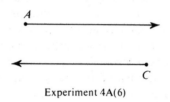

Experiment 4A(6)

7 Draw a pair of intersecting segments. Consider them as the diagonals of a quadrilateral by joining their endpoints. Is there anything special about this quadrilateral? What?

8 Repeat Experiment 4A(7), making the diagonals the same length. Note the figure formed.

9 Repeat Experiment 4A(7), making the diagonals the same length and having only one bisect the other. Note the figure formed.

10 Repeat Experiment 4A(7), making the diagonals the same length and having both diagonals bisect each other. Note the figure formed.

11 Repeat Experiment 4A(7), making the diagonals of different lengths and having them bisect each other. Note the figure formed.

12 Repeat Experiment 4A(7), making the diagonals of different lengths and perpendicular bisectors of each other. Note the figure formed.

13 Repeat Experiment 4A(7), making the diagonals the same length and perpendicular bisectors of each other. Note the figure formed.

14 See if you can formulate new definitions for square, rectangle, rhombus, and parallelogram on the basis of Experiments 4A(7) through 4A(13).

98

EXERCISE 4A

1 Construct quadrilaterals (using compass and straightedge) having exactly one, two, and four right angles. Why can't a quadrilateral have exactly three right angles?

2 Is it possible for a parallelogram to have exactly one pair of right angles? Two pairs?

3 Prove deductively that each of the three conditions for a figure to be a parallelogram yields the other two. (Hint: Draw a parallelogram, construct a diagonal, and use congruent triangles.)

4 Determine which of the figures must be parallelograms based on the given information.

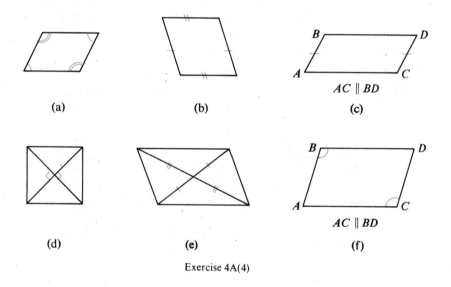

(a)   (b)   (c)

(d)   (e)   (f)

Exercise 4A(4)

5 Why is it sufficient to have only one right angle in a parallelogram to ensure that it is a rectangle?

6 Is a square a parallelogram?

7 Is it enough to define a square as a quadrilateral with one pair of adjacent sides congruent?

8 Is a rhombus a square? Is a square a rhombus? Is a trapezoid a rhombus?

99

9 Using your knowledge of interior angles, explain why a construction of type (a) is rigid while type (b) will sway or fall.

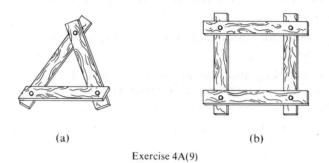

(a)   (b)

Exercise 4A(9)

10 What must be done to structures of type (b) to ensure stability? Examine the framework of a construction project, a steel tower, or a similar structure.

11 Prove that the diagonal of a parallelogram divides it into two congruent triangles, using the definitions in this section.

12 Prove that all the diagonals of a parallelogram bisect each other.

13 In the illustration $ABCD$ is a parallelogram with $\overline{BF} \perp \overline{AD}$ and $\overline{ED} \perp \overline{BC}$. Prove $\triangle ABF \cong \triangle EDC$.

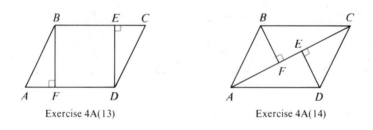

Exercise 4A(13)   Exercise 4A(14)

14 In the illustration $ABCD$ is a parallelogram with $\overline{BF} \perp \overline{AC}$ and $\overline{DE} \perp \overline{AC}$. Prove $\overline{BF} \cong \overline{DE}$.

4.2 REGULAR GEOMETRIC FIGURES

A geometric figure with interesting properties is the *regular* figure, in either two or three dimensions. The following definition is for two dimensions.

Definition 4.1 A *regular polygon* is one in which all sides are congruent to each other and all angles are congruent to each other.

100

Some regular polygons are commonly found in designs and patterns: the hexagon in floor tile and inlays, the octagon in the stop sign. We have already seen examples of regular triangles and quadrilaterals; the equilateral triangle and the square fit the requirements of Definition 4.1. Construction of regular polygons with more than four sides is complicated, and sometimes impossible, using a straightedge and compass. The key to the construction of these figures can be found in the angles shown at the center of the polygon in Illus. 4.9. For example, a regular hexagon

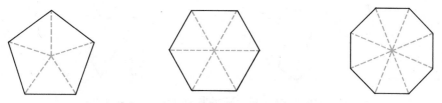

ILLUSTRATION 4.9   Regular polygons

can be formed from six congruent equilateral triangles; this should suggest a method of construction. The regular octagon has opposite pairs of sides parallel, and each side is perpendicular to the pair of sides just beyond the adjacent sides. This also should suggest a method for construction.

In addition to the regular polygons, of which there is an unlimited number, there are five regular polyhedra.

Definition 4.2   A *regular polyhedron* is a convex polyhedron whose faces are determined by congruent regular polygons, and whose dihedral angles are all congruent.

Since it has been some time since we discussed polyhedra, we remind you that in a three-dimensional figure we are interested in the surface, including the interiors of the polygons as well as the polygons themselves. The regular polyhedra and the patterns for their construction are pictured in Illus. 4.10 and 4.11. The faces must be triangular, square, or pentagonal. A regular polyhedron cannot have a higher-order regular polygon for a face.

ILLUSTRATION 4.10   Regular polyhedra

101

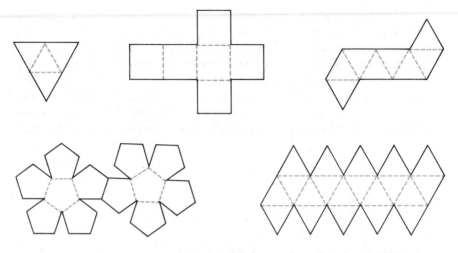

ILLUSTRATION 4.11   Models for regular polyhedra

EXPERIMENT 4B

1   Sketch regular polygons of six, eight, ten, and twenty sides. What geometric figure does the series of regular polygons approach in shape as the number of sides increases? Does this suggest a method for describing that figure?

2   Develop, or look back at earlier sections for, methods of constructing regular polygons of three, four, six, and eight sides.

3   Can you find a relationship between straight angles and the union of the angles of a regular hexagon? Pentagon?

4   See if you can determine why there are only five regular polyhedra by considering Experiment 4B(3) above and the following questions:
   (a) How many regular triangles may have common vertices and thus form the vertex of any polyhedron?
   (b) How many squares may share a common vertex and thus create a vertex of any polyhedron?
   (c) How many regular pentagons may share a common vertex and thus create a vertex of any polyhedron?
   (d) How many regular hexagons may share a common vertex and thus create a vertex for any polyhedron?
   (e) Give a reason why it is of no value to pursue the quest for regular polyhedra further?

5   Using the patterns suggested in the text (or your own) construct, of stiff paper or cardboard, each of the regular polyhedra.

6   Illustration 4.11 gives patterns for each of the regular polyhedra. Find at least one other pattern for each of them.

102

7 The intersection of two faces of a polyhedron is an *edge*. The intersection of three or more edges is a *vertex* of the polyhedron. Calculate the possible number of edges and vertices for the given number of faces. There is more than one possibility for polyhedra with more than four faces.

Faces	Edges	Vertices	Name
4	?	?	tetrahedron
5	?	?	pentahedron
6	?	?	hexahedron
7	?	?	heptahedron
8	?	?	octahedron
10	?	?	decahedron

Do you see any relationship among the columns?

EXERCISE 4B

1 Give three examples of each of these regular polygons in the world around you: triangle, square, pentagon, hexagon, octagon. Can you find one example of each figure occurring as a work of nature and not of man?

2 We found in the experiments that if V, E, and F represent the numbers of vertices, edges, and faces of a simple polyhedron, then $V - E + F = 2$. This is known as Euler's formula. Apply Euler's formula to the five regular polyhedra to determine the number of edges of each.

3 Must the faces of regular polyhedra be regular polygons? Must these polygons be congruent?

4 How many different regular polyhedra can be formed using squares as faces? Using pentagons? Using hexagons? Why?

5 How many faces does a regular polyhedron have if the faces are equilateral triangles such that three meet at each vertex of the polyhedron?

6 How many faces does a regular polyhedron have if the faces are equilateral triangles such that four meet at each vertex of the polyhedron?

7 How many faces does a regular polyhedron have if the faces are equilateral triangles such that five meet at each vertex of the polyhedron?

8 If two congruent regular tetrahedra are placed together to form a new polyhedron, is the new polyhedron regular? Explain.

9 Look up a method in a college geometry textbook for the construction of a regular pentagon.

4.3 CIRCLES

From a mathematical point of view one of the most interesting geometric figures is the circle. This figure holds the attention of the artist, the naturalist, and the physicist. It is one of the simplest figures to construct

103

with a compass, yet the relationship of its diameter to its circumference is one of the most sophisticated in elementary mathematics. This particular relationship eluded mathematicians for centuries.

We shall adhere to our technique of defining figures as sets of points and define a circle in the following way:

Definition 4.3   Given in one plane a point O called the *center* and a segment $\overline{OA}$ called the *radius*, a *circle* C is the set of all points X such that $\overline{OX}$ is congruent to $\overline{OA}$; that is, $C = \{X \mid \overline{OX} \cong \overline{OA}\}$.

This definition, based on the idea of congruence of line segments, has many immediate results. One of these is that *all radii of the same circle are congruent*. In Illus. 4.12, $\overline{OA}$ and $\overline{OE}$ are radii of the circle. They both

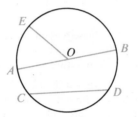

ILLUSTRATION 4.12   A circle with diameter $\overline{AB}$ and chord $\overline{CD}$

have one endpoint at the center O and they are congruent segments. It also follows that two circles with congruent radii are congruent to each other. If two circles have the same center but do not have congruent radii, they are called *concentric* circles.

Consider two points A and B on a circle (see Illus. 4.12) such that $\overline{OA}$ and $\overline{OB}$ are radii and A, O, and B are collinear; $\overline{AB}$ is a *diameter* of the circle. Any segment with endpoints on the circle (such as $\overline{CD}$ in Illus. 4.12) is a *chord* of the circle. The diameter $\overline{AB}$ is thus a chord of the circle. If two chords having a common endpoint as in Illus. 4.13(a) are

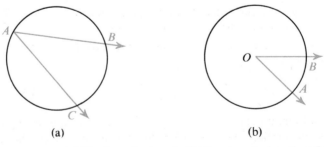

(a)   (b)

ILLUSTRATION 4.13   Inscribed angle CAB and central angle AOB

104

extended to form rays, they form an angle which is said to be *inscribed* in the circle. Two radii in the same circle, when extended, form an angle called a *central angle* (Illus. 4.13(b)).

If a line intersects a circle, the segment determined by the circle is a chord. The line itself is a *secant*. If the line intersects the circle at one and only one point as in Illus. 4.14, the line is called a *tangent* to the circle. A radius drawn to the point of tangency is perpendicular to the tangent, and is the shortest distance from the center to the tangent.

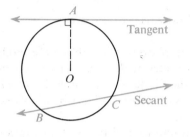

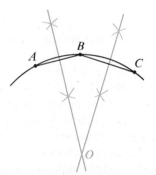

ILLUSTRATION 4.14 A tangent and a secant of a circle

ILLUSTRATION 4.15 Reconstructing a circle from a piece of arc

An *arc* of a circle is any connected subset of the curve forming the circle. An arc is denoted by its endpoints: in Illus. 4.14 the arc from B to C is denoted $\overset{\frown}{BC}$. If two points are given, such as B and C, then two arcs are formed: a major arc $\overset{\frown}{BC}$ and a minor arc $\overset{\frown}{BC}$. A circle is bisected by a diameter into two congruent arcs; each of these is termed a *semicircle*. It is possible to construct a circle if any portion of the arc is given. As shown in Illus. 4.15, we locate three convenient points A, B, and C on the arc and draw the segments $\overline{AB}$ and $\overline{BC}$. Using compass and straightedge we construct the perpendicular bisectors of $\overline{AB}$ and $\overline{BC}$. The intersection of the bisectors, labelled O in the illustration, is the center of the circle. Since any point on the perpendicular bisector of a segment is equidistant from the endpoints, O is equidistant from A, B, and C. It follows then that O is equidistant from any point on the arc. This technique for finding the center of a circle is used for rebuilding wheels. It is possible to reconstruct any wheel if a portion of the wheel is available.

If a circle is enclosed by tangent lines so that the segments of the tangents form a polygon as in Illus. 4.16(a), the circle is said to be *in-*

105

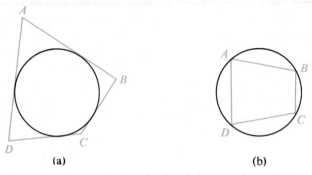

ILLUSTRATION 4.16   Inscribed and circumscribed circles

scribed in the polygon, or the polygon is *circumscribed* about the circle. On the other hand, if a polygon is constructed with all its vertices on the circle, then the polygon is *inscribed* in the circle, or the circle is *circumscribed* about the polygon. It is not always possible to inscribe a polygon in a circle or inscribe a circle in a polygon.

Many properties of circles can be found by a little experimentation. As an example, suppose we draw a circle as has been done in Illus. 4.17 and then draw some chord $\overline{AB}$ of the circle. Now construct a radius of the circle perpendicular to $\overline{AB}$: that is, construct a perpendicular to $\overline{AB}$ through the point O. Using a compass we compare the segments $\overline{AC}$ and $\overline{CB}$. It appears that these segments are congruent. Was this congruence just an accident or would the same be true for any chord? To help us decide we could draw several more chords and repeat the experiment, and in each case the result would be the same. Using inductive reasoning we would surmise that the radius perpendicular to a chord always bisects the chord.

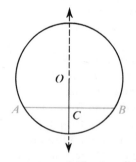

ILLUSTRATION 4.17   A radius perpendicular to a chord

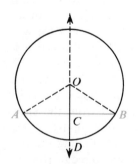

ILLUSTRATION 4.18   A line passing through O perpendicular to a chord bisects the chord

Since induction has led us to what we think is a valid conclusion, let us try to establish the conclusion deductively.

THEOREM 4.1 *If a line through the center of a circle is perpendicular to a chord, it bisects the chord.*

We will use Illus. 4.18 as we develop the proof. It shows a chord $\overline{AB}$ and a line $\overrightarrow{OD}$ perpendicular to $\overline{AB}$ at C. Since $\overrightarrow{OD}$ is perpendicular to $\overline{AB}$, $\angle ACO \cong \angle BCO$, and they are both right angles. $\overline{AO} \cong \overline{BO}$ as these segments are both radii of the same circle, and $\overline{OC} \cong \overline{OC}$, consequently $\triangle AOC \cong \triangle BOC$ and therefore $\overline{AC} \cong \overline{BC}$. This proves the theorem deductively. Several more theorems will be developed both inductively and deductively in the exercises.

The circle is defined in a plane. The counterpart of a circle in three dimensions is the *sphere*.

Definition 4.4   Given a point O in space and a line segment $\overline{AB}$, then the set $\{X \mid \overline{OX} \cong \overline{AB}\}$, where X is a point in space, is called a *sphere*.

You should recognize that this definition describes a surface. As for a circle, we say the radius of the sphere is the segment $\overline{AB}$ in Definition 4.4. Alternately we can say the radius is any line segment joining the center and a point of the sphere. Similarly, the diameter of a sphere is a line segment through the center with its endpoints on the sphere.

The definitions of chord, secant, and tangent apply as well to spheres as to circles by substituting the word "sphere" for the word "circle" in each definition. In addition, we can describe a plane as being tangent to a sphere if it has one and only one point in common with the sphere. If a plane intersects a sphere in more than one point, then the intersection is always a circle. The largest of all such circles, which has as its center the center of the sphere, is called a *great circle* (Illus. 4.19). Note that if a plane and sphere intersect in more than one point, then they intersect in

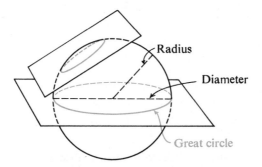

ILLUSTRATION 4.19   A radius and diameter of a sphere

107

an infinity of points. Illustration 4.16 shows a circle inscribed in a polygon and a polygon inscribed in a circle. In much the same way certain polyhedra can have spheres inscribed in them or circumscribed about them. A ball placed inside a box so that all sides of the box, as well as the top and bottom, touch the ball is a physical example of a sphere inscribed in a polyhedron.

The sphere is one of nature's favorite forms and is extremely useful to man. As well as having appeal to the eye, it has strength of shape and economy of material. That is, a sphere encloses the maximum amount of space within a minimum surface area and does so with a form very resistant to pressure. It has the additional property that it will roll in any direction. These physical properties are all manifestations of its definition. We see the ability to roll freely in ball bearings, the economy of material in balloons and bubbles, and the strength in the shape of some water storage tanks.

EXPERIMENT 4C

1 Draw a circle and compare the diameter to any other chord of the circle. Are they congruent? Which seems larger? Can you draw any conclusions?

2 Inscribe an angle in a semicircle. What kind of angle does it seem to be? Repeat this experiment, placing the vertex in several positions. Does this seem to make any difference.

3 Construct two congruent chords in the same circle. Then construct perpendiculars from the center of the circle to each of the chords. Compare the lengths of the perpendiculars. Any conclusions?

4 Construct a central angle in a circle. Complete a triangle by connecting the ends of the radii. What kind of triangle is this?

5 Extend a radius $\overline{AO}$ to C to form a diameter as shown. Draw the chord $\overline{CB}$ and the radius $\overline{OB}$. What kind of triangle is OCB?

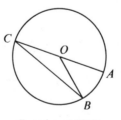

Experiment 4C(5)

6 Compare the central angle AOB to the inscribed angle BCA. How are they related in measure?

108

7 With a compass set to the length of the radius, mark off an arc $\overset{\frown}{AB}$ on a circle. Again using the radius, mark off an arc $\overset{\frown}{BC}$. Repeat the process until you have come back to A. Does it appear to "come out even"?

8 Use a circle to develop an alternate means of constructing the following regular figures:
(a) Regular triangle   (b) Regular quadrilateral
(c) Regular hexagon   (d) Regular octagon

9 What comparison between the diameter and circumference of a circle can you make using an inscribed equilateral triangle? An inscribed regular hexagon? Other regular polygons?

10 Construct two concentric circles. Can you draw a line tangent to both? Why, or why not?

11 Draw a circle and two unequal chords in the circle. Which chord is closer to the center of the circle? Is your answer always the same?

12 Approximate the intersection of two planes by folding a piece of cardboard along a line. Place two unequal spherical beads in the dihedral angle formed. What conjectures can you make relating the diameters of the beads and the distances from the fold to the points of tangency?

13 Draw a circle. Through a point A outside the circle draw two tangents to the circle. Compare the segments from A to the points of tangency. Are they congruent?

14 Draw a circle and a diameter of the circle. Construct a line tangent to the circle at each end of the diameter. How are these tangents related?

15 Place a coin flat on your desk and arrange two pencils so that they are tangent to the coin and intersect. What do you know about the line segment joining the center of the coin to the point of intersection?

16 Construct a triangle. Locate the center of the circle that inscribes the triangle. What previous construction (Experiment 3D) uses the same process?

17 Construct a triangle. Locate the center of the circle inscribed in the triangle. What previous construction (Experiment 3C) used the same process?

18 Construct a right triangle. Inscribe the triangle in a circle. Where is the center of the circle in relation to the triangle? Repeat for several other right triangles.

19 Draw a convenient segment $\overline{AB}$ representing one side of a square, and construct the square. Now inscribe the square in a circle.

20 Draw convenient segments $\overline{AB}$ and $\overline{CD}$ representing the length and width of a rectangle, and construct the rectangle. Now inscribe the rectangle in a circle.

21 Draw a circle with radius congruent to $\overline{AB}$ in Experiment 4C(20) and center X. Using X as a center draw another circle with radius congruent to $\overline{CD}$, where $(\overline{CD} \ncong \overline{AB})$. How would you relate the two circles?

109

22 Draw a circle with radius $\overline{AB}$ and center X. Select two points C and D on the circle such that $\overline{CD}$ is not a diameter of the circle. Draw $\overline{XC}$ and $\overline{XD}$. Are $\overline{XC}$ and $\overline{XD}$ congruent?

23 Construct a circle and label the center O. Mark two points on the circle, A and B. Draw $\overline{OA}$ and $\overline{OB}$. How do their lengths compare?

24 In Experiment 4C(23) what figure is formed by $\overline{OA} \cup \overline{OB} \cup \overline{BA}$?

25 Mark two points X and Y exterior to the circle in the drawing for Experiment 4C(23). Draw $\overline{XY}$. Does $\overline{XY}$ intersect the circle? Does it intersect exactly once? Exactly twice? Does your answer depend on the position of X and Y?

26 In Experiment 4C(23) can you draw another circle, distinct from the first, that goes through A, B, and some third point C on the first circle?

EXERCISE 4C

1 What is the least information you can have about two circles and know they are congruent?

2 What is the least information you can have about two spheres and know they are congruent?

3 Prove that two congruent chords of a circle are equidistant from the center.

4 Prove that the tangents to a circle from a point outside the circle are equal.

5 Prove that $m(\angle AOB) = 2m(\angle ACB)$ using the illustration.

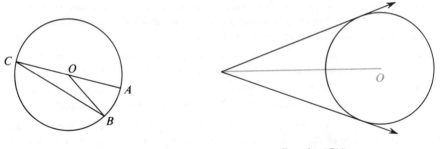

Exercise 4C(5)   Exercise 4C(6)

6 Prove that the bisector of the angle formed by two tangents passes through the center of the circle in the illustration.

7 Is it possible to inscribe a sphere in every regular polyhedron? Give reasons.

8 What figure is formed by the intersection of two spheres?

9 How many small circles can be passed through a point on a sphere? How many great circles?

110

10 How many small circles can be passed through two points on a sphere not the endpoints of a diameter? How many great circles?

11 Display all the possibilities for the intersection of a line and a circle. A plane and a sphere.

12 Consider two planes that intersect a sphere and have congruent circles as their intersections. Are the planes equidistant from the center of the sphere?

13 Can a circle have two centers? A sphere?

Before going further we summarize as theorems a few of the more important ideas from the experiments and exercises.

THEOREM 4.2 *In the same circle or in congruent circles equal chords are equidistant from the center.*

THEOREM 4.3 *In the same circle or congruent circles if two chords are unequal, the longer chord is closer to the center.*

THEOREM 4.4 *If a line is tangent to a circle it is perpendicular to the radius drawn to the point of tangency.*

THEOREM 4.5 *The segments formed by tangents to a circle through a point outside the circle are congruent.*

THEOREM 4.6 *An angle inscribed in a semicircle is a right angle.*

4.4 PRISMS AND OTHER THREE-DIMENSIONAL FIGURES

In Section 4.2 we studied a few of the regular geometric figures in three dimensions. In this section we shall examine several categories of geometric solids, including polyhedra, and observe some of their more important properties.

Prisms are a special class of polyhedron in which at least two faces are congruent polygons. Some examples are cubes, boxes, buildings, rooms, and many packages. It is easiest to define a prism if we first discuss a surface called a *closed prismatic surface*.

Let *P* be a polygon in a plane (Illus. 4.20) and let *L* be a line in space that intersects the plane somewhere on the polygon. Consider the set of all lines parallel to line *L* that also intersect the polygon. The union of all these lines is a *closed prismatic surface*, and each line is an *element* of the surface. There are many physical counterparts of this kind of surface; for example a shower curtain like that in Illus. 4.21 can be thought of as a surface generated by all the lines passing through the polygonal rod at the top. With this understanding of a closed prismatic surface we can define a prism.

111

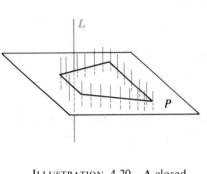

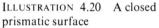

ILLUSTRATION 4.20 A closed prismatic surface

ILLUSTRATION 4.21 A physical example of a closed prismatic surface

Definition 4.5 A *prism* is a solid cut off from a closed prismatic surface by two parallel planes which cut all the elements of the surface.

Illustration 4.22 shows a prism. The faces of the sections made by the two parallel planes are called the *bases* of the prism. Each of the

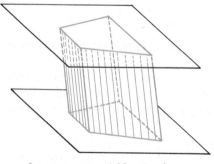

ILLUSTRATION 4.22 A prism

faces of the prismatic surface between the two parallel bases is a *lateral face* of the prism. In Illus. 4.22 there are four lateral faces. Several features of prisms are immediately obvious. From the definition and observation, the bases are congruent and parallel. The lateral *edges*, that is the intersections of the lateral faces, are also parallel; hence the lateral faces are all parallelograms. If the lateral edges are perpendicular to the bases, then the lateral faces are rectangles, and the prism is known as a *right* prism. Prisms are generally named by the shape of the polygonal base. Thus the prism in Illus. 4.22 is a quadrilateral prism while the ones

112

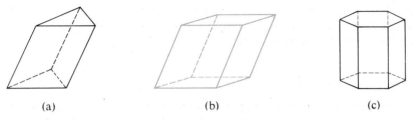

(a)   (b)   (c)

ILLUSTRATION 4.23   Triangular, quadrilateral, and hexagonal prisms

shown in Illus. 4.23 are (a) triangular, (b) square, and (c) hexagonal prisms. More specifically, Illus. 4.23(c) is a right hexagonal prism.

A prism with parallelograms for its bases *and* for its lateral faces is known as a *parallelepiped*. An example of a parallelepiped can be seen in a deck of cards in a stack (Illus. 4.24). Even if the deck is sloped, the

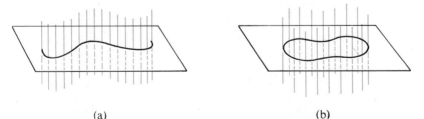

ILLUSTRATION 4.24   Parallelepipeds represented by stacked layers

figure is a parallelepiped. Special cases of the parallelepiped are rectangular prisms and right rectangular prisms. The familiar rectangular solid with all faces perpendicular to adjacent faces and all vertices formed by three perpendicular lines is a right rectangular solid. A *cube* is a rectangular parallelepiped with all edges congruent.

Other three-dimensional figures can be generated in much the same manner as the prism. Rather than generating a closed prismatic surface, we generate a *cylindrical surface* by finding the union of all lines (again called elements), parallel to a given line, that intersect some curve in a plane. Such a figure is depicted in Illus. 4.25(a). A sheet of paper held

(a)   (b)

ILLUSTRATION 4.25   Cylindrical surfaces

113

upright on the surface of a desk is a realization of this surface. If the curve in the plane is a simple closed curve (Illus. 4.25(b)), the resulting figure is a *closed cylindrical surface*.

Definition 4.6   A *cylinder* is a solid cut off from a closed cylindrical surface by two parallel planes cutting all the elements of the surface.

Illustration 4.26 shows three cylinders. As you can see, Definition 4.6 includes a greater variety of shapes than a layman might include in his definition of a cylinder.

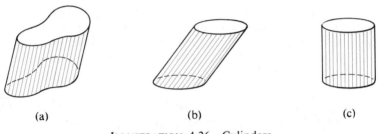

(a)　　　　　　　　　(b)　　　　　　　　　(c)

ILLUSTRATION 4.26   Cylinders

If the elements of the cylindrical surface are perpendicular to the base, the cylinder is a *right cylinder*. If the simple closed curve in the plane is a circle, the resulting figure is a *circular cylinder*. The right circular cylinder is of course, the shape used for tomato-soup cans, oatmeal boxes, silos, and railroad tank cars. Tubes, pipes, and conveyer belts are physical examples of cylindrical surfaces without bases.

Another figure related to the prism is the *pyramid*. The union of all lines (elements) that intersect a given polygon and also intersect a fixed point Q not in the plane of the polygon (Illus. 4.27) determines a *pyra-*

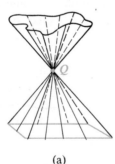

(a)　　　　　　　　　　　　　(b)

ILLUSTRATION 4.27   Pyramidal surfaces

114

midal surface. The point Q is called the *vertex* of the surface. You can see in Illus. 4.27 that the union of all elements along one side of the polygon generates a surface that lies within a plane. You can also see that a pyramidal surface has two parts intersecting at the vertex. These parts are called *nappes* of the surface. The figure we commonly call a pyramid can now be defined.

Definition 4.7 A *pyramid* is a solid cut from one nappe of a pyramidal surface by a plane cutting all its elements.

Pyramids are usually named by the shape of the base. Thus there are triangular, rectangular, and hexagonal pyramids, among others. Illustration 4.28 shows examples of the first two types.

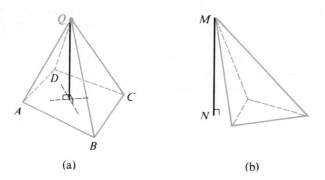

(a) (b)

ILLUSTRATION 4.28   Pyramids with altitudes

If the base of a pyramid is a regular polygon, we can locate a point interior to the polygon which is equidistant from all the vertices. If the segment from the vertex to this center point is perpendicular to the base, as is the one in Illus. 4.28(a), the pyramid is called a *right pyramid*. The segment from the vertex to the center of the base of a right regular pyramid is called the *axis* of the pyramid.

In any case the line segment from the vertex of a pyramid perpendicular to the base (or to the plane of the base) is an *altitude* of the pyramid. Illustration 4.28(b) shows that the altitude may lie outside the pyramid. The lateral surfaces of any pyramid are triangles.

A figure having the same relationship to the pyramid as the cylinder does to the prism is the *cone*. If the generating curve is not polygonal then the surface is a conical surface rather than a pyramidal surface.

Definition 4.8 A *cone* is a solid cut from one nappe of a conical surface by a plane cutting all its elements.

115

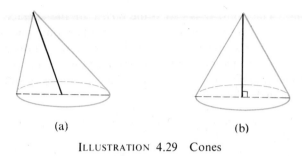

ILLUSTRATION 4.29   Cones

The cone has a curved lateral surface and a planar base. For most people, the word "cone" nearly always conjures up the *right circular cone* shown in Illus. 4.29(b); such a cone has a circular base and its altitude must be perpendicular to the base at its center. We see these figures in many contexts, from party hats to ice cream cones to rays from spotlights.

A modification of the pyramidal or conical shape made by slicing the top off the figure with a plane parallel to the base is called a *frustum* of a pyramid or cone. Wastepaper cans and lampshades are usually frustums of cones. The pedestal of a statue may be the frustum of a pyramid. In these cases the figure has two bases and a lateral surface. Illustration 4.30 shows these basic shapes.

ILLUSTRATION 4.30   Frustums of a cone and a pyramid

EXPERIMENT 4D

1 The figure shows a parallelepiped and a diagonal plane *B'BDD'*. How many other diagonal planes can be drawn in this figure? Is *AA'C'C* one of them? Sketch it in the drawing.

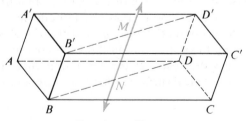

Experiment 4D(1)

116

2 In Experiment 4D(1) the two diagonal planes $B'BD'D$ and $AA'C'C$ intersect in a line $\overleftrightarrow{MN}$. What line segments appear to be bisected by each other?

3 Sketch a right circular cylinder. Imagine slicing it with a plane parallel to the base. What is the shape of the figure in the plane of the slice?

4 In Experiment 4D(3), suppose the slice is not parallel to the base, but still intersects all of the curved surface. Now what figure results?

5 Sketch a noncircular cylinder. Can a cross section of this cylinder be a circle? How?

6 Consider the lateral faces of the regular pentagonal pyramid shown. Are they all congruent triangles? How do you know? Are they isosceles triangles?

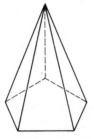

Experiment 4D(6)

7 Sketch a right circular cone. Intersect it with a plane through its vertex and through its base. Intersect it with a plane parallel to the base and not through the vertex. Describe these intersections. What other intersections are possible?

8 Cut a rectangle 3 inches by 5 inches of cardboard. Draw a line along one 5-inch edge. Stand it upright on your desk so that the 5-inch edge is perpendicular to the plane of the desk, and pivot it around on an axis through the unmarked 5-inch edge. What figure is described by the marked edge? Can you formulate a definition for a cylinder from your observations?

9 Cut a right triangle from cardboard. Place it upright on your desk so that one side of the right angle is perpendicular to the plane of the desk. Pivot it around one leg while holding the one leg in the plane of the desk. What figure is described by the hypotenuse?

10 Tie five broken rubber bands together with a single knot. Draw a pentagon and pin the loose end of a rubber band at each vertex. Pull the bands taut. Can you see the pyramid formed? Can the knot be moved to different positions and still result in the formation of a pyramid?

EXERCISE 4D

1 Which of the five regular polyhedra are prisms? Pyramids? Rectangular solids? Parallelepiped?

117

2 Consider the set of all congruent circles which have their centers located on a given circle, where the radius of the given circle is greater than the radius of the congruent circles. What set of points in space does this constitute?

3 Find a formula relating the total number of faces of a prism to the number of sides of the base polygon from which it was generated?

4 What is the least number of faces a prism can have? The greatest?

5 Can a cross section of a pyramid ever be a trapezoid? Why?

6 How many edges does a pentagonal pyramid have?

7 Find a formula relating the total number of edges of a pyramid to the number of sides of the base polygon from which it was generated?

8 If the center of a cube is connected with each vertex of the cube, how many pyramids are formed?

9 Is the frustum of a pyramid a polyhedron?

10 Can the frustum of a pyramid be a regular polyhedron?

11 A feed trough for poultry can be an example of a frustum of a pyramid. Name two others.

12 Is it possible for a section or slice across a noncircular cylinder, made parallel to the base, to be circular?

13 What is the shape of the cross section of a right circular cone sliced by a plane perpendicular to the axis?

4.5 SIMILARITY AND PROPORTIONS

A relation closely allied to congruence is that of *similarity*. In fact, we can describe similarity as congruence of shape (angles) without regard to congruence of size (length). Physical examples of similar figures are numerous. Road maps are similar to the actual roads, while men's shirts in different sizes, scale models, and photographic enlargements are examples of similar figures. First, we define similarity of geometric figures composed of line segments.

Definition 4.9 Two geometric figures composed solely of line segments are *similar* if and only if their corresponding angles are congruent and their corresponding sides are proportional. If figures *A* and *B* are similar, we write $A \sim B$.

This definition allows for figures of one, two, or three dimensions to be similar. Some figures such as circles do not have measurable angles. Their similarity depends on mathematics beyond our scope, but it should be obvious that all circles are similar to each other.

Note that any two line segments are similar. Also any two congruent

118

figures are similar. Congruency implies similarity, but the converse is not true.

In examining the geometric properties of similar figures, we find that the motion of similarity is closely related to the parallel postulate. To show this relationship, consider two triangles as shown in Illus. 4.31.

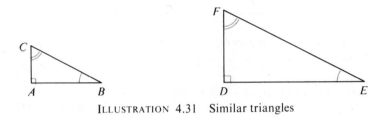

ILLUSTRATION 4.31   Similar triangles

For the two triangles to be similar, their corresponding angles must be congruent. Since the union of the three angles of a triangle is congruent to a straight angle, it follows that if two angles of one triangle are congruent to two angles of a second triangle, the third angles must also be congruent. Consequently if two angles of one triangle are congruent to two angles of a second, the triangles must be similar. Now place the two similar triangles in Illus. 4.31 in the position shown in Illus. 4.32, where *A*

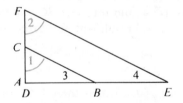

ILLUSTRATION 4.32   Similar triangles positioned to show parallelism

and *D* coincide as do *D-C-F* and *A-B-E*; then $\overline{CB}$ falls in the interior of $\triangle DEF$. Since $\angle 1 \cong \angle 2$, the segments $\overline{CB}$ and $\overline{FE}$ must be parallel.

The converse is also true. If we are given $\triangle DEF$ with $\overline{CB} \parallel \overline{FE}$ then $\angle 1 \cong \angle 2$, as these are corresponding angles of parallel lines cut by a transversal; $\angle D \cong \angle A$ by identity, and hence the two triangles are similar.

We do not need congruency of sides of triangles to have similarity of figures, yet similarity of figures does imply a relation between corresponding sides of the two figures. If we are given any triangle such as $\triangle ABC$ in Illus. 4.33, and bisect $\overline{AB}$ at *D* with a compass and then con-

119

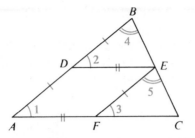

ILLUSTRATION 4.33   A line parallel to a side of a triangle bisects both of the other sides if it bisects one of them

struct $\overline{DE} \parallel \overline{AC}$ and then at E construct $\overline{EF} \parallel \overline{AB}$, we see that $\overline{AD} \simeq \overline{DB}$, $\measuredangle 1 \simeq \measuredangle 2 \simeq \measuredangle 3$; and $\measuredangle 4 \simeq \measuredangle 5$ as shown in the figure. $ADEF$ is a parallelogram by definition; thus $\overline{AD} \simeq \overline{FE}$ and $\overline{AF} \simeq \overline{DE}$.

$\triangle BDE \simeq \triangle EFC$ by the angle-side-angle theorem, so $\overline{BE} \simeq \overline{EC}$ and $\overline{DE} \simeq \overline{FC}$. Together this means E and F must also be midpoints of the sides of triangle ABC.

This says, as an immediate conclusion, that a line parallel to one side of a triangle and bisecting a second must also bisect the third side. We can also say that a line parallel to one side of the triangle bisects both of the other sides if it bisects one of them.

The implications of this result are extensive. We could have located a point on $\overline{AB}$ one-fourth of the way from A to B and would have found that a line parallel to $\overline{AC}$ would intersect $\overline{BC}$ one-fourth of the way from C to B. We state the general result without further discussion.

THEOREM 4.7   *The segments intercepted on any two transversals by three or more parallel lines are proportional to each other.*

In Illus. 4.34, $L_1 \parallel L_2 \parallel L_3$; if B is three-fifths of the way from A to C then the theorem states that E is three-fifths of the way from D to F.

An important conclusion obtained from this theorem is that the sides

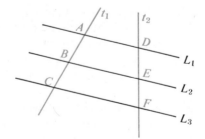

ILLUSTRATION 4.34   Proportional segments on parallel lines cut by transversals

120

of similar polygons are proportional. It is this result that enables us to construct scale models or to make drawings using projected and parallel lines.

By locating P beyond $\triangle ABC$ in Illus. 4.35, and drawing the projection lines L_1 and L_2 we can draw a larger model of the triangle by drawing $\overline{A'B'} \parallel \overline{AB}$ and $\overline{B'C'} \parallel \overline{BC}$.

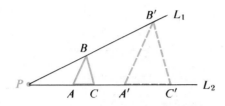

ILLUSTRATION 4.35   Using a projection for enlarging

We can also use a variation of proportional segments to divide a given line segment into any number of congruent parts. Suppose we wish to divide a segment such as $\overline{AB}$ in Illus. 4.36 into five congruent parts. At

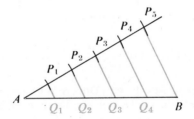

ILLUSTRATION 4.36   Dividing a line segment into five congruent parts

A we construct a line segment at a convenient angle to $\overline{AB}$. Setting our compass at a convenient length, we mark off five consecutive congruent segments on this line. We draw $\overline{P_5B}$ and then construct lines through P_1, P_2, P_3, and P_4 parallel to $\overline{P_5B}$. By the proportional property we know that $\overline{AQ_1} \cong \overline{Q_1Q_2} \cong \overline{Q_2Q_3} \cong \overline{Q_3Q_4} \cong \overline{Q_4B}$. Hence $\overline{AB}$ is divided as required.

EXPERIMENT 4E

1   Construct any triangle, locate the midpoints of the three sides, and join them with segments. This forms four small triangles. Compare these four triangles for congruency. Repeat for at least one other triangle.

121

2 Compare the larger triangle to one of the smaller triangles in Experiment 4E(1). Are they similar?

3 Draw a triangle ABC like that in Illus. 4.33 with $DE \parallel AC$. Is $\triangle BDE$ similar to $\triangle ABC$? Can you give a deductive proof?

4 Angle B is a right angle and $\overline{BD} \perp \overline{AC}$. What triangles are similar?

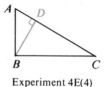

Experiment 4E(4)

5 In the diagram $\overline{DC} \parallel \overline{AB}$. Is $\triangle DOC$ similar to $\triangle AOB$? Can you give a deductive proof?

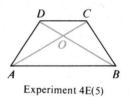

Experiment 4E(5)

6 In Experiment 4E(5), is $\triangle ADB$ similar to $\triangle ACB$? Can you give a deductive proof?

7 Consider the three figures shown. Does A appear to be similar to B? Does B appear to be similar to C? Does it follow that A is similar to C?

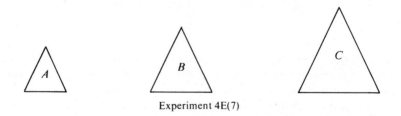

Experiment 4E(7)

8 Divide this line segment into three congruent sections. Into ten congruent sections. Was the "convenient unit" on the auxiliary line the same in both cases?

Experiment 4E(8)

9 This figure has been superimposed on a square grid. Use graph paper with ¼-inch squares to draw an enlargement. Is your figure similar to this one?

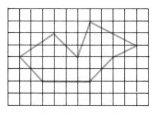

Experiment 4E(9)

10 Use ¼-inch graph paper to draw an enlargement of the figure. Is your result similar to this one? Can you answer with the same degree of certainty as in Experiment 4E(9)?

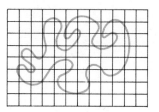

Experiment 4E(10)

EXERCISE 4E

1 Are all regular polygons of a given number of sides similar? All regular polyhedra of a given number of faces?

2 If a movie projector is projecting an image of a triangle onto a screen so that the axis of projection is perpendicular to the screen, will the angles in the triangle on the screen be congruent to those in the triangle on the film? Why? Are the figures similar?

3 If the projector in Exercise 4E(2) projects the image onto the screen obliquely, will the image be similar to the figure on the film?

4 Prove the theorem: A line joining the two midpoints of two sides of a triangle is parallel to the third side.

5 Which of the following classes of figures are all similar to each other within the class?
   (a) Rectangles   (b) Cubes   (c) Spheres
   (d) Pyramids   (e) Rhombuses   (f) Cylinders
   (g) Right triangles

123

6 Deduce which triangles are similar in the figure.

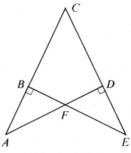

Exercise 4E(6)

7 Name pairs of similar figures in the rectangle.

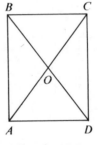

Exercise 4E(7)

8 List five examples of commercial products that are similar.

9 Discuss the similarity of the image on a 23-inch TV screen to that on an 11-inch screen. What about the similarity of the image to the actual scene being televised?

10 Are all equilateral triangles similar? All cones? All regular hexagonal pyramids?

11 How many angles of two quadrilaterals must be known to be congruent if you are to state that the quadrilaterals are similar?

12 For any three polygons *A*, *B*, and *C*, is *A* similar to *A*? If *A* is similar to *B*, is *B* similar to *A*? If *A* is similar to *B* and *B* is similar to *C*, is *A* similar to *C*? What kind of a relation is "is similar to" on the set of polygons?

4.6 TRANSFORMATIONS: MOTION GEOMETRY

In the geometry we have studied thus far, we have assumed a given fixed set of points and defined relations on this set. In none of the work covered

124

have we suggested that we might try to move points from one location to another. In earlier work we were able to compare segments for the relation "less than" by using a set of dividers or a compass to carry the measurements from one segment to another. It is quite important to see that we did not try to move an endpoint of the first segment and superimpose it on an endpoint of the second segment to establish the relation, but rather used a physical device to establish the relation. Intuitively, however, you must sense that the relation could be established this way. The geometry of motion is certainly a natural process encountered by children as they deal with physical models. A child covers one plywood triangle with another to see if they fit, that is, to establish congruence. In much the same manner a seamstress slides a pattern onto material to cut a dress of a given size.

Motion geometry, or tranformation geometry as it is more often called, requires a different set of assumptions than those we have been using. In our study of motion geometry we will think in terms of physical models of geometric figures and study several motion operations. When mathematicians study geometry they do not think in terms of physical sets but rather think of abstract sets and an abstract correspondence between sets. We will refer to this abstract approach only casually.

The geometry of transformations establishes a correspondence between two figures using some kind of matching or *mapping* from one figure to the other. Thus, the rubber band in Illus. 4.37 can be made to

Rubber band

Steel wire

ILLUSTRATION 4.37   Establishing correspondence in geometric figures

correspond to the circle or the square by moving it to a suitable position and reshaping. Even a non-stretchable substance such as a steel wire can be moved to the circle or square position. However, the rubber band can be stretched to encompass the larger square whereas the wire can not. This is another way of saying that the circumference of the steel wire, the given circle, and the smaller square are fixed or *invariant* while the circumference of the rubber band is not invariant. Although we cannot stretch the steel wire from the small square shape to the large square

125

shape, each point on the small square can be corresponded to a point on the large square and vice versa; that is, we can establish a one-to-one correspondence between the points on the small square and those on the large one. The following is an intuitive definition of transformation that will serve for our purposes.

Definition 4.10   A transformation is an operation that produces a one-to-one correspondence between the points of two geometric figures.

In transformation geometry we are concerned with properties of figures that remain invariant under transformations or motions. Scientists are very concerned with the invariance of physical properties under physical transformations. For example, the weight or mass of a given amount of water remains unchanged when the water (in the form of ice) is heated from 32°F to 100°F but the form of the material does change. We say that the mass is invariant under the given conditions, whereas the form is variant.

We will study the behavior of geometric figures under various motions to see what properties remain unchanged. Many transformations are possible, but we will limit our discussion here to (1) translation, (2) rotation, (3) reflection, and (4) similarity.

(1)   *Translation.*   The translation motion is the simplest of all transformations in geometry; it involves a shift or slide in the position of the figure. Illustration 4.38 shows two translations that can be made

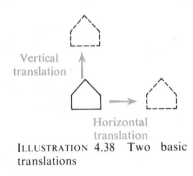

Vertical translation

Horizontal translation

ILLUSTRATION 4.38   Two basic translations

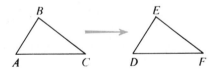

ILLUSTRATION 4.39   Two triangles to test for congruence

in two dimensions. A test for congruence of triangles resting side by side in the plane as shown in Illus. 4.39 is to translate or slide one on top of the other to see if they coincide point for point. If there is point-for-point correspondence then the two figures are congruent.

In the case of triangles we need only consider the point-for-point correspondence of the three vertices (that is, only three points) as all other points are determined by the vertices.

In Illus. 4.40, △*ABC* can be made to coincide with △*DEF* by making

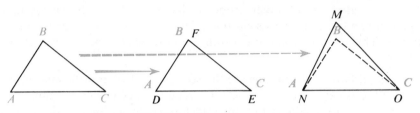

ILLUSTRATION 4.40   Using translation to establish congruence

the vertex at *A* correspond with that at *D*, the vertex at *B* correspond with that at *F*, and the vertex at *C* correspond with that at *E*. However, △*ABC* does not coincide with △*MNO* by any kind of translation. No matter how we slide it vertically or horizontally we cannot get point-for-point correspondence between the two triangles. Vertex *A* can be made to correspond with *N* and *C* with *O*, but then *B* will not correspond with *M*.

In Illus. 4.41 we encounter a different problem. The rectangle *ABCD*

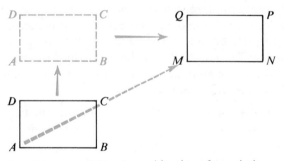

ILLUSTRATION 4.41   A combination of translations

appears to be congruent to the figure *MNPQ*. However, a vertical or horizontal slide alone does not place *ABCD* on *MNPQ*. The problem is easily solved, as the reader has already guessed, by a combination of two translations, one vertical and one horizontal, or by a single translation along $\overline{AM}$. In three dimensions we can use a combination of three translations to move a given object from one position to another.

Many objects moved from one position to another, such as boxes in a warehouse, people down the corridors of a building, or a train down a straight subway passage, are examples of translation. Remember that a translation involves a *linear* shift only. What invariant properties are there to a figure that is subjected to a translation? Are its shape, area, altitude, angles, or any other property changed? The only change in a figure after a translation is its *position* relative to a fixed scale or reference frame. This is why we can use a translation to determine congruence of figures.

(2) *Rotation*. Rotations are the second kind of transformation we will consider. A rotation, or turn, requires a center of rotation. All other points of the figure are transformed to another position in a circular arc about this point. The two triangles shown in Illus. 4.42

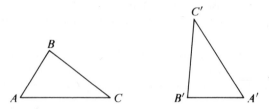

ILLUSTRATION 4.42   Two triangles that are not congruent by translation

are congruent, but if we slide the first triangle onto the second so that *A* coincides with *B'* the other vertices will not coincide. There is no way to translate or slide △*ABC* onto △*A'B'C'* to get coincidence of these points. To get △*ABC* in the coinciding position we must revolve or rotate it about some fixed point, *A* for example, and then use a horizontal translation.

Illustration 4.43 shows one way this rotation can be accomplished. Rotate △*ABC* about *A* as a center, then translate △*ABC* horizontally to

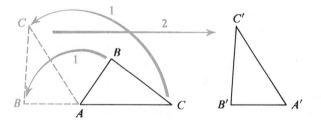

ILLUSTRATION 4.43   A combination of rotation and translation establishes congruence

128

the right. (Of course, there are other combinations of rotation and translation that would result in coincidence of the two figures.) Curiously, there is a unique point K between the two triangles such that a rotation about K will bring the two triangles into coincidence—no translations are necessary.

Rotation of a figure also leaves many properties of the figure unchanged. Its area, congruence, size, etc., all remain constant. The change brought about by a rotation is in relative position. A line that was horizontal before rotation will, in general, not be horizontal after rotation.

Some of the definitions we made in the earlier part of the chapter can be restated in a different but equivalent way using what we know about transformations. For example, a square was defined as a special case of the rectangle, one which has two adjacent sides congruent. Under the transformation of rotation we can see that for a rectangle to coincide with itself under one-fourth of a complete rotation about its center it must have adjacent sides congruent. Other figures can be defined as figures that coincide with themselves under different kinds of rotation; Illus. 4.44(b) shows coincidence of a parallelogram with itself under one-half of a complete rotation.

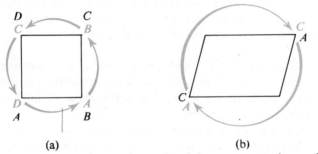

(a)   (b)

ILLUSTRATION 4.44 Rotation can be used to define a square and a parallelogram

(3) *Reflection.* . The two triangles in Illus. 4.45 show still another problem with reference to congruence that we would like to resolve

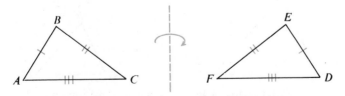

ILLUSTRATION 4.45 Congruence by reflection

129

by transformations. We would say the triangles are congruent by corresponding the parts as shown, yet there is no way to translate or rotate $\triangle ABC$ to make it coincide with $\triangle DEF$. These triangles are reversed from left to right. The only way to get coincidence of corresponding sides $\overline{AB}$ with $\overline{DE}$, $\overline{BC}$ with $\overline{EF}$, and $\overline{AC}$ with $\overline{DF}$ would be to reverse the image of $\triangle ABC$ (or $\triangle DEF$) by rotating it out of the plane of the paper about a vertical axis (that is, the perpendicular bisector of $\overline{CF}$) and replacing it in the plane; then we would have the desired correspondences. Such a transformation is called a *reflection in the plane about a line*. More informally we describe this motion as a *flip*.

This transformation also preserves shape and size but the figure reflected loses the property of direction or left to right order.

The three transformations, reflection, translation, and rotation, all preserve congruence of figures as invariant properties and are thus called *rigid-motion* transformations. They are sometimes called *Euclidean transformations* because they can be accurately represented on paper with the classic tools of Euclidean geometry.

(4) *Similarity.* The similarity transformation is closely related to the concept of figure similarity discussed earlier in this chapter. We can intuitively think of a similarity transformation as one that changes the size but not the shape of a figure.

Every housewife who has had a garment shrink in hot water understands the implications of this kind of transformation. Similarity as a transformation plays an important role in the realm of image reproduction—photography, printing, etc.

Scale drawings are, essentially, similiarity transformations. In such cases we are concerned with the ratio of the transformation. In Illus. 4.46, if the two triangles are similar, then we can show a correspondence between segments determined by pairs of points, such as $\overline{AB}$ and $\overline{A'B'}$.

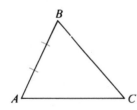 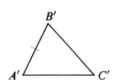

ILLUSTRATION 4.46   A similarity transformation

The correspondence must be the same *regardless of the points chosen*; it is called the *ratio of similarity*. In Illus. 4.46 the ratio of similarity is 3:2; that is, the ratio 3:2 transforms $\triangle ABC$ into $\triangle A'B'C'$. A similarity ratio of 1:1 denotes congruence.

EXPERIMENT 4F

1 Draw a dashed line segment on a sheet of paper. Is this congruent to itself under translation? Rotation? Reflection? Similarity?

2 Draw an isosceles triangle. Can you make it similar to itself under translation? Rotation? Reflection? Similarity?

3 Repeat Experiment 4F(2) for a scalene triangle.

4 Consider an equilateral triangle. What part of a rotation is required before it is congruent to itself?

5 Repeat Experiment 4F(4) for a square, a regular pentagon, and a regular hexagon, then try to generalize to a definition for any regular polygon.

6 See if you can develop alternative definitions for the following figures using the ideas of transformations. There may be more than one.
   (a) Segment   (b) Circle
   (c) Right triangle   (d) Angle
   (e) Isosceles trapezoid   (f) Rectangle

7 Draw a circle on a sheet of paper. Does it coincide with itself under translation? Rotation? Reflection?

8 Consider the figure. Give a set of translation directions to move (a) so it coincides with (b). Can you give another set of directions to obtain the same goal?

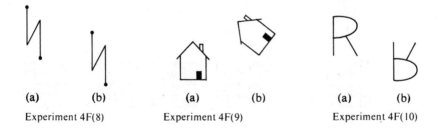

(a)   (b)   (a)   (b)   (a)   (b)

Experiment 4F(8)   Experiment 4F(9)   Experiment 4F(10)

9 Give a set of translation and rotation directions to move (a) to (b). Can you find another set of directions to accomplish the same thing?

10 Give a set of translation and rotation instructions to move (a) to (b). Do you encounter a problem? What additional transformation is needed?

131

EXERCISE 4F

1 What kind of transformations are involved in the following situations? Indicate how many transformations are used and the order of their use.

(a) A picture is moved from one wall to an adjacent wall in a room.

(b) An elevator rises from the fourth floor to the ninth floor.

(c) A page is turned in a book.

(d) An automobile is backed out of a garage and reparked in the garage rear first.

(e) The hour hand of a clock changes from a reading of 3 to 4.

(f) A bowling ball is thrown down an alley.

(g) A balloon is blown up from a smaller to a larger size.

(h) A snapshot is enlarged.

(i) A knob on an air conditioner is turned from HI to LOW.

2 Is it possible for a sequence of two rotations to be equivalent to a single translation? If so, give an example.

3 What geometric figures in the plane are congruent to themselves under any translation?

4 What geometric figures in the plane coincide with themselves under any rotation?

5 Can a similarity transformation produce a figure that is congruent to itself? If so, how?

6 In each case determine which of the figures on the right can be obtained from the figure on the left by translation, rotation, and reflection.

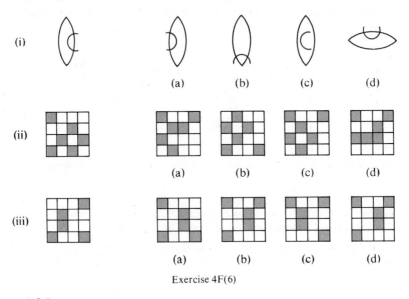

(i)   (a)   (b)   (c)   (d)

(ii)   (a)   (b)   (c)   (d)

(iii)   (a)   (b)   (c)   (d)

Exercise 4F(6)

132

7 Give directions using translations and rotations to get each figure through the door. Can you remove all the figures?

Exercise 4F(7)

8 The only figure which is a pure reflection of *A* is ____.

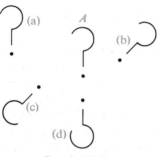

Exercise 4F(8)

9 Which of the following are correct? Figure 1 is the image of figure 2 under
(a) A turn followed by a slide
(b) A turn followed by a flip
(c) A slide followed by a flip
(d) All of these

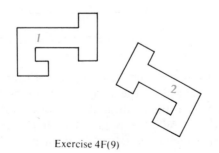

Exercise 4F(9)

133

10 In each case decide if *A* is congruent to *B*.

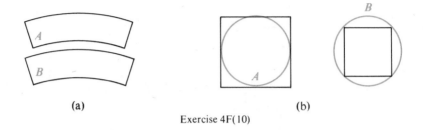

(a)

(b)

Exercise 4F(10)

4.7 SYMMETRY

A property of geometric figures that has been in evidence in some of the cases we have examined, but remains to be defined, is the property of *symmetry*. There are two kinds of symmetry in the plane and three kinds of space symmetry.

Examine a pair of gloves (or your hands). Are they congruent? There certainly is a relationship between the two objects. Yet no matter how we place the two figures, one never coincides with the other. You are well aware that two gloves lack congruence in three dimensions if you try to interchange them on your hands, yet their shapes are the same. The problem is that one is the image of the other in a reversed direction. This property of having the same shape but reversed image is *symmetry*.

Let us consider symmetry in two dimensions first. A figure such as the isosceles triangle shown in Illus. 4.47 is balanced about the dashed

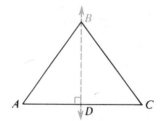

ILLUSTRATION 4.47 A line of symmetry

line $\overleftrightarrow{BD}$. By drawing this line we observe that triangle *ABD* is congruent to triangle *BCD*, although the images are reversed. We cannot leave △*ABD* and △*BDC* on the plane and slide one over the other to get co-incidence of points. What is required is that we pick one triangle up, *flip it over in space about line $\overleftrightarrow{BD}$* and return it to the plane (a reflection

134

transformation). Whenever this leads to congruence we say that the plane figure is *symmetric about a line*, and that the line is an *axis of symmetry*.

We can define line symmetry of a plane figure as follows:

Definition 4.11 Given a plane figure and a line. Draw a perpendicular to the line through a point P on the line, intersecting the figure at X_1 and X_2. If $\overline{X_1 P} \simeq \overline{X_2 P}$ for every point P on the line, then the figure is symmetric about the line.

Illustration 4.48 shows the application of Definition 4.11 to (a) a line-symmetric figure, and (b) a figure in which $\overline{X_1 P} \not\simeq \overline{X_2 P}$ and hence in which

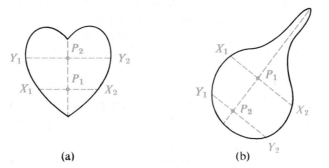

(a)   (b)

ILLUSTRATION 4.48   The heart is symmetric about a line but the other figure is not

line symmetry does not exist. Note that no matter where P is picked on the axis in (a), the segments $\overline{X_1 P}$ and $\overline{X_2 P}$ are congruent. In (b) it may be possible to find some congruent segments, but if even one pair of segments are not congruent, the figure is not symmetric.

This definition of symmetry makes use of the reflection transformation. A physical test for symmetry is to fold a paper figure along the assumed axis of symmetry: if the two portions match, the figure is symmetric.

Illustration 4.49 shows several figures with their axes of symmetry.

The second type of plane symmetry is symmetry about a point, which can be defined as follows:

Definition 4.12 Given a plane figure F and a point P. If for every point $X_1 \in F$ we can find another point $X_2 \in F$, such that $\overline{X_1 P} \simeq \overline{X_2 P}$ and $X_1 - P - X_2$, then the figure is said to be *symmetric* about the *point*. The point P is called a *center* of symmetry.

135

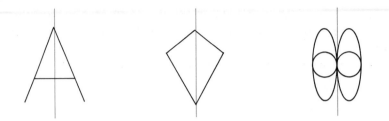

ILLUSTRATION 4.49   Three examples of figures with line symmetry

This definition says that for every point on the figure, there is another point on the figure the same distance away from the center and on the same line. Thus we can rotate or turn a point-symmetric figure about the center of symmetry and get coincidence of the figure with itself in one or more positions *other than* the original position. For example, the block letter N has point symmetry. Revolved half-way around (upside down), the letter is coincident with itself. Illustration 4.50 shows some

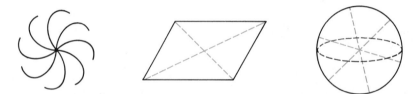

ILLUSTRATION 4.50   Point symmetry in two- and three-space

examples of point symmetry. Note that point symmetry and line sym-metry do not imply each other, though a figure may have both (check the line-symmetric figures in Illus. 4.49 for point symmetry). A circle is an example of the latter case; it is symmetric about its center and about every diameter.

Symmetry in three-space is harder to describe and our discussion will be intuitive.

Point symmetry in three-space involves an extension in all directions of point symmetry in two dimensions. Each point of the figure must have an image on the opposite side of the center, a corresponding distance away. A sphere, a rectangular parallelepiped, and a dumbbell are ex-amples of figures with three-dimensional point symmetry.

Line symmetry in three-space requires point symmetry in every plane perpendicular to the axis of symmetry and line symmetry in every plane

136

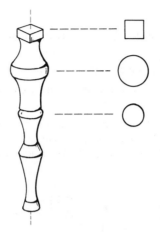

ILLUSTRATION 4.51   A turned table leg displays symmetry

containing the axis of symmetry. The ornate table leg in Illus. 4.51 has three-dimensional line symmetry. Every plane section (three are shown) has point symmetry, and every plane containing the axis of the leg has line symmetry about that axis.

The base of a lamp, the wheel of a car, and a washer all have line symmetry. A test of line symmetry is to ask, "Will it spin smoothly on this axis." Three-dimensional line symmetry does not imply three-dimensional point symmetry (as in Illus. 4.51) but both may be possible in the same figure.

Plane symmetry is peculiar to three-dimensional figures and demands that any point on a given figure have a corresponding point on the opposite side of a plane of symmetry, the same distance from the plane along a line perpendiculr to the plane. This can be interpreted in the physical sense as the figure's being able to be sliced by a plane so that the two parts are left-hand and right-hand images of each other in three dimensions.

Just as we could not make the left and right halves of an equilateral triangle coincide as long as we left them in the plane, so we cannot demonstrate plane congruence of three-dimensional objects (such as gloves) as long as we leave them in three-space. If we had a fourth dimension to pick the objects up through, to turn them over, and then replace them in three-space, we could demonstrate the congruence of such figures. The impossibility of doing this does not prevent us from saying that a right-hand glove and a left-hand glove of the same pair are congruent images of each other through a reflection about a plane.

137

EXPERIMENT 4G

1   Construct a line segment. Is it symmetric about an endpoint? Any point on itself?

2   Consider a line. Is it symmetric about a point on itself? Every point?

3   Can a plane figure have more than one axis of symmetry? Construct examples.

4   Fold down the top half of a large sheet of paper. Fold the paper in half the other way. Draw a line segment from corner to corner as shown, and cut along the segment. Unfold the sheet. What figure do you have? How many lines of symmetry can you find?

Experiment 4G(4)

5   Repeat Experiment 4G(4) starting with a square piece of paper. Now how many lines of symmetry are there?

6   Refold the cutout in Experiment 4G(4) along its folds. Cut several small pieces out of the folded edges. Unfold the paper. Does this figure have the same lines of symmetry as the figure in Experiment 4G(4)?

7   Draw a curve on paper. Place a mirror perpendicular to the paper so that the image is reflected in the mirror and "touches" the curve. Do the image and the curve form a symmetric figure?

8   The accompanying sketch is of a house. Certain features are missing that would make the house symmetric about a plane. What else would you add to complete the drawing to obtain symmetry? Why?

Experiment 4G(8)

9   Draw a figure which has point symmetry in the plane. Line symmetry in the plane. Both.

138

10 Determine the line of symmetry for this figure. What previously discussed properties do you use?

Experiment 4G(10)

EXERCISE 4G

1 Determine the types of symmetry that exist in each of the following:

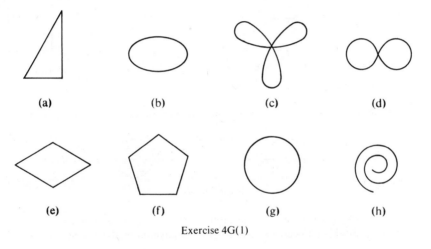

(a)   (b)   (c)   (d)

(e)   (f)   (g)   (h)

Exercise 4G(1)

2 Do two planes of symmetry result in line symmetry? Point symmetry in three dimensions?

3 Decide if each of the following figures is symmetric about the axis or point dictated.
(a) A square about its diagonal
(b) A square about one of its sides
(c) A pinwheel about its hub
(d) A circle about its diameter
(e) A four-leaf clover about its stem juncture
(f) An automobile body about a vertical plane along the lengthwise axis
(g) An airplane about a vertical plane through the axis of the fuselage
(h) A parallelogram about one of its diagonals

139

4 Tell if the figures below are symmetric with respect to the axes specified.

   (a) $\overline{AD}$ (b) $\overline{BE}$ (c) G

   (d) $\overline{KL}$ (e) $\overline{JM}$ (f) J

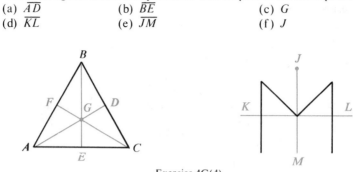

Exercise 4G(4)

5 Find the point of symmetry in each of the following:

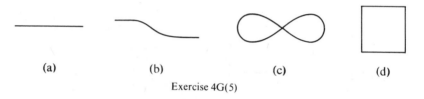

   (a)   (b)   (c)   (d)

Exercise 4G(5)

6 Determine if the figure below is symmetric about the axis indicated.

Exercise 4G(6)

7 Which of the 26 letters of the alphabet have axes of symmetry? Consider both capital and lowercase letters. Do any have symmetry about a point?

8 Determine which regular geometric figures of less than 20 sides have centers of symmetry. Do you see any pattern developing so that you can make a general statement?

9 Which of the regular polyhedra have planes of symmetry?

10 Which of the following pieces of sports equipment have axes of symmetry? Centers of symmetry? Planes of symmetry?

   (a) Tennis racquet   (b) Football

   (c) Basketball court   (d) Basketball goal

   (e) Basketball   (f) Baseball

   (g) Baseball bat   (h) Baseball diamond

   (i) Hockey rink   (j) Hockey club

   (k) Ice skates   (l) Golf club

140

11 List five objects that *grow* that are symmetrical (or nearly so) about a line, point, or plane.

12 The word CHECK written in block style capital letters is symmetric about a horizontal line passing through the crossbar of the H. Think of several other words with this property.

13 The words MOM and TOOT written in block style capital letters are symmetric about a vertical line. Name two other common words with this property.

14 If the letters of the word TOMATO are written in block style capital letters one under the other, the result is symmetric about a vertical line through the centers of the letters. Name two other words with this property.

15 Complete each of the figures so that they are symmetric about the given axis.

(a)   (b)   (c)

Exercise 4G(15)

16 If a figure has two axes of symmetry, must the figure be symmetric about the point of intersection of the two figures?

17 Consider a figure that has two planes of symmetry. Must the figure be symmetric about the line of intersection of the two planes?

18 What kind of symmetry is apparent in many women's dresses? Name two articles of apparel that are not symmetric about a line or a plane.

19 How many lines of symmetry does a five-pointed star have? A six-pointed star? An eight-pointed star? Can you generalize?

20 What relation exists between point symmetry and rotation? Between line symmetry and reflection?

141

5 / GEOMETRY WITH MEASURE

5.1 IDEAS BEHIND MEASURE

In previous chapters we developed geometric ideas and relationships without, for the most part, using the concept of measure of area, volume, or length. This may seem unusual, since most of the practical applications of geometry involve some sort of measure. But by developing the notion of congruence first, we have laid the foundation for discussion of systems of measure, such as the English system used in this country.

Someone in a remote part of the Earth, who knows nothing about measure, could decide for himself whether or not two sticks are the same length simply by comparing them side by side. He would exemplify determination of congruence without a measuring system. But a much more sophisticated system is needed for a person to decide if the refrigerator-freezer unit in the showroom several miles from his home will fit into the niche (or go through the door) in his kitchen. Similarly, the structural ironworker needs to make use of more advanced concepts of congruence to guarantee that the beam he is about to position by crane will exactly fit the space between two columns.

Several assumptions must be made before we can discuss the technical aspects of measurement in geometry. Measurement is a physical application of a geometric concept; hence the assumptions are more physical than mathematical in nature.

Postulate 5.1   There exists a one-to-one correspondence between the set of real numbers and the points on a line.

We suppose that the reader is familiar with the set of real numbers and the concept of line. Postulate 5.1 is the basis for all linear measure as well as the starting point of an approach to geometry called *analytic geometry* (see Chapter 8).

By Postulate 5.1, we can associate a real number with each point of a line, and associate exactly one point on the line with each real number. We can depict the association as follows:

On any line, label some handy reference point with the real number 0. To the right of this point, locate another point and associate the real number 1 with it (Illus. 5.1). Construct a second segment congruent to the

ILLUSTRATION 5.1   The points 0 and 1 on the number line

segment $\overline{01}$ with its left endpoint at 1. Associate the number 2 with the right endpoint of the new segment. Continue this process both to the right and to the left (for negative integers) until a number line similar to that in Illus. 5.2 is formed. This process associates each integer with a point on the number line.

ILLUSTRATION 5.2   Integral points on the number line

It is imperative that the segments constructed so far be congruent if the number line is to be used for measure. For other applications, this may not be necessary; a slide rule is one example in which non-congruence is required.

These correspondences on the number line embody the first postulate, but also make use of another (unstated) assumption that was inherent in our work in the last chapter:

Postulate 5.2   If line segments $\overline{AB}$ and $\overline{CD}$ are congruent in one position, then they are congruent in any other position in which they may be placed.

Thus we assume that the length of a line does not change simply because it is moved. Another way of stating Postulate 5.2 is to say that a segment is invariant under the three transformations translation, rotation, and reflection. If we use a ruler to measure a board and also to measure the place into which the board is to fit, the measurements will not be distorted by the motion of the ruler. Imagine the confusion if this were not true. Suppose all lines on the surface of the Earth going in a

143

north-south direction became longer (or shorter) when they were turned
to an east-west direction!

*Postulate 5.3   A given line segment may be subdivided into as many sub-
segments as desired.*

We can now subdivide the number line in Illus. 5.2 to locate points
on it in as many places as we choose. These subdivisions provide points
associated with the fractions (Illus. 5.3 depicts some of them), and by ex-

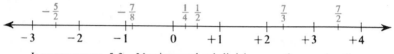

ILLUSTRATION 5.3   Nonintegral subdivisions on the number line

tension, all the rational numbers. It might seem that the process of corresponding fractions with points on the line would eventually cover the
line completely. However, this is not the case; no matter how many
rational numbers are located, there still remain unallocated points. Numbers like $\sqrt{2}$ and $\sqrt{3}$ are not rational, yet there are points on the number
line that correspond to them.

The $\sqrt{2}$ is known to correspond to the hypotenuse of a right triangle with equal arms. (This will be discussed in Section 5.4; it is known
as the Pythagorean relationship.) A line of length $\sqrt{2}$ constructed on the
number line, as in Illus. 5.4, shows that there is a point on the line cor-

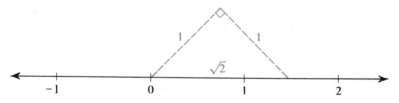

ILLUSTRATION 5.4 $\sqrt{2}$ corresponds to a point on the number line

responding to $\sqrt{2}$. Points corresponding to $\sqrt{3}$ and $\sqrt{5}$ and all other
irrational numbers can be located by similar methods.

We can now define length of a line segment.

*Definition 5.1   Given any line segment $\overline{AB}$ with real numbers a and b
associated with the points A and B, two differences $a - b$ and $b - a$ can
be formed. Whichever of these differences is positive is the length of $\overline{AB}$,
also referred to as the measure of $\overline{AB}$. We symbolize this measure as
$m(\overline{AB})$.*

144

The definition requires that we choose $b - a$ or $a - b$ to represent the measure of $\overline{AB}$, whichever is positive. Mathematicians use the symbol $|b - a|$ to signify the difference $b - a$ without regard to sign; this is called the *absolute value* of $b - a$. With this terminology, the definition becomes $m(\overline{AB}) = |b - a|$.

We can now measure along a straight line, once we have an appropriate unit of measure. Of practical importance is the fact that, for example, it is possible to conclude that the new refrigerator $32\frac{1}{2}$ inches wide and 58 inches high will go through the kitchen door 36 inches wide and 75 inches high even though the appliance and the door are miles apart.

EXPERIMENT 5A

1 Using a standard ruler (divided into sixteenths) measure the distance between the two points, starting at A and measuring to B. Record your measurement. Then measure the distance between the two points starting at B and measuring to A. Was this measurement the same as the first? Repeat this experiment for two other points.

Experiment 5A(1)

2 In the illustration, if a is the point on the number line which corresponds to 1 and b is the point which corresponds to $2\frac{1}{2}$, what is the value of $b - a$? What is the value of $a - b$? Are you two answers the same? Are they related? What is the value of $|a - b|$? Of $|b - a|$?

Experiment 5A(2)

3 Locate two points A and B on a number line. Measure the distance from A to B. Measure the distance from B to A. Compare your results. What is the distance from A to A? From B to B? Compare your results.

4 Locate three points on a number line and label them A, B, and C in order. Measure the distance from A to B. Measure the distance from B to C. Measure the distance from A to C. How does the last number compare with each of the other two? With the total of the other two?

5 Using an eye estimate, which of the two segments is longer: $\overline{AB}$ or $\overline{CD}$? Now measure with a ruler.

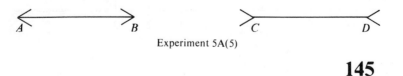

Experiment 5A(5)

145

EXERCISE 5A

We denote the line segment from A to B by $\overline{AB}$, and the measure of the line segment by $m(\overline{AB})$. Complete each of the first four statements using the results of the experiments. These constitute the basic properties of normal measure.

1 For any two points A and B on the number line, $m(\overline{AB})$ _____ $m(\overline{BA})$.

2 For any point A on the number line, $m(\overline{AA})$ = _____.

3 For any three points A, B, and C on the number line such that A-B-C, $m(\overline{AB}) + m(\overline{BC}) = m(\underline{\quad})$.

4 The distance between any two points A and B on the number line can be denoted by the formula _____.

5 If it is true that in a right triangle $c = \sqrt{a^2 + b^2}$, construct line segments equal in length to $\sqrt{3}$, $\sqrt{5}$, and $\sqrt{10}$.

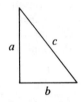

Exercise 5A(5)

6 List three everyday experiences that use the idea that the distance from A to B is the same as the distance from B to A. Then see if you can explain why a westbound nonstop jet flight from New York to San Francisco takes longer than one going eastbound over the same route at the same airspeed.

7 What postulate is commonly violated when a fisherman describes "the one that got away?"

8 Shown here is a logarithmic scale. Which postulate does it violate that makes it unfit for ordinary measure?

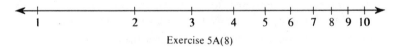

Exercise 5A(8)

9 Make a sketch of a measuring stick that doesn't conform to Postulate 5.1. Could you measure anything with it? If so, what?

10 Find the distance between A and B, C and D, E and F on the scale below.

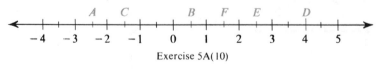

Exercise 5A(10)

11 Using Definition 5.1, find the lengths of $\overline{AB}, \overline{AC}, \overline{BC}, \overline{AD}, \overline{AE}, \overline{CE},$ and $\overline{DE}$.

Exercise 5A(11)

12 What is $|-5|$? $|6|$? $|0|$?

5.2 LINEAR MEASURE

The three postulates of the last section refer to measure in a straight line, or *linear measure*. Similarly, the experiments and exercises were concerned with linear measure. To measure distance in a line we established a convenient unit between integral points and made measurements in terms of multiples of this basic unit. The choice of the basic unit is arbitrary and usually is dictated by the relative size of the objects to be measured. In ancient times, units of measure were poorly defined. The Bible mentions the *cubit*, described as "the distance from the elbow to the tip of the finger." Obviously this unit of distance varied from person to person. Much later the English described a unit called the inch as the "length of three barleycorns laid end to end" and the yard as "the distance from the King's nose to the tip of his outstretched hand." Today, however, the definitions of units such as foot, yard, and meter are precisely stated. Each of these units is convenient for measuring on many applications, but other units are used for extremely small or extremely large measures in technical work. An astronomer uses the *light year* or *parsec* as a standard unit to measure interstellar distances, and the laboratory physicist uses the *angstrom* as a basic unit for measuring microscopic lengths. Two different *systems* of measure, involving two different units, are in use today. These are the English and the metric systems of measure, to be discussed more fully in Chapter 7.

Let us suppose we are given a handy unit of length, such as that shown in Illus. 5.5, and wish to determine the distance between *A* and *B* in terms of that unit. We shall call the basic unit a *handy*. In order to establish the distance between *A* and *B* in terms of handys we must make use of a principle first declared by the Ancient Greeks and attributed to Archimedes.

Postulate 5.4 Given two quantities (or numbers) x and y, and y greater than x, then there exists a finite number n such that the statement $nx \geq y$ is true.

147

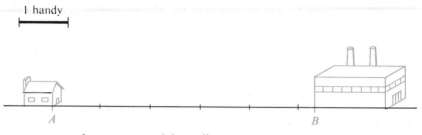

ILLUSTRATION 5.5   A distance measured in handys

In a measuring situation we can interpret this postulate along with Postulate 5.1 to mean that it is possible to measure the distance between any two points, no matter where they are located, using any given segment as the standard unit. This is to say, the distance from Denver to Chicago could be measured in miles or in inches, although in the latter case n would be an extremely large number.

In Illus. 5.5, the distance from A to B can be expressed in terms of handys by applying this principle. Notice that the distance is not equal to an integral number of handys, but is somewhere between 5 and 6 handys. By using the idea of subdivision we can divide the basic unit into fourths and measure the entire distance as $5\frac{1}{4}$ handys.

One rule of thumb concerning measure is to use a unit that is convenient. Thus we use inches, feet, yards, miles, etc., for different measuring tasks. But each of these units is related to the others to make them all more meaningful and useful measuring tools. Thus, 12 inch units are combined to make 1 foot unit, and the foot is extended to create a 3-foot unit and a 5280-foot unit.

Thus far we have considered linear measure only along a line, yet we all have applied the "straight" units to measure a curve. The length of any simple closed curve is called its *circumference*, and the sum of the lengths of the sides of a polygon is called its *perimeter*. Both these measures are given in linear units. The justification for applying linear measure to nonlinear objects involves some difficult theoretical mathematics. The calculation of the actual circumference of a circle with a diameter of 5 units (the diameter is linear) involves mathematics well beyond the scope of our work here. However, we can approximate a curve length using intuitively acceptable techniques.

Consider the curve in Illus. 5.6. Its length can be found with a reasonable degree of accuracy by considering several points on the curve between A and B. Suppose we consider four interior points as marked in Illus. 5.7. Then the length of the curve is approximated by the sum of the lengths of the segments $\overline{AP_1}$, $\overline{P_1P_2}$, $\overline{P_2P_3}$, $\overline{P_3P_4}$, and $\overline{P_4B}$. We are approxi-

148

ILLUSTRATION 5.6   A curved line joins *A* and *B*

mating the measure of a non-polygonal curve by using the measure of a polygonal curve. If the approximation is not sufficiently accurate, a better measure can be obtained by choosing more points and measuring more, but shorter, segments. Thus, the length of the curve from *A* to *B* is more closely approximated by the polygonal curve of 13 segments in Illus. 5.8.

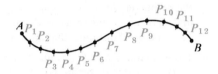

ILLUSTRATION 5.7   The   length of the curve joining *A* and *B* can be approximated by five segments

ILLUSTRATION 5.8   Thirteen segments gives a better approximation of the length

This idea of using more and more segments of shorter length to approximate a curve can be carried to any desired degree of accuracy. In actual application, however, the practical solution is to use a measuring tool made of flexible material so that we do not have to subdivide the curve into segments. Thus, the distance around a circular pillar is best measured by a flexible steel tape. If the only available measuring device were a rigid ruler, it would be necessary to use the subdividing technique.

EXPERIMENT 5B

1   Use a ruler to measure a piece of rope 12 inches long. Form a circle with the rope. Is it farther around on the inside or on the outside of the circle? How do you explain this?

2   In Experiment 5B(1), what could be done to reduce the difference in the two lengths?

3   What is your conclusion about the accuracy of using a tape measure to measure around an object? What could improve its accuracy?

4   Measure the diameter of a circle. Approximate the length of the circumference of the circle by taking one-half the diameter and stepping off this length around the circle with a compass. Repeat using one-fourth, one-eighth, and one-sixteenth of the diameter as the measuring segment. What values did you obtain? How many times the diameter is the circumference?

149

5 Suggest a method for obtaining a good approximation to the length of the circumference of a circle, as a result of Experiment 5B(4). Try your method.

6 Cut a circle from lightweight cardboard. Mark some point on the edge of the cardboard. Hold the circle on end, with the marked point against a paper on which you have drawn a long line segment. Put a mark on the line segment opposite the mark on the cardboard. Roll the circle along the line segment until the mark once again touches the paper. Put a second mark on the line segment at this point. Measure the diameter of the cutout and the marked line segment. Compute the ratio of length of line segment to diameter. Is the ratio approximately 22:7?

7 Approximate the circumference of the circle of Experiment 5B(4) by inscribing it in a square, an octagon, and a 16-sided polygon and measuring the perimeters of the polygons. Then circumscribe the circle about a square and an octagon, and measure the perimeters of these polygons. Can you suggest a range of distances that includes the true length of the circle?

8 Draw a circle. Use the radius to mark congruent segments on it. Can you mark six segments before going around the entire circumference? Can you mark seven segments before passing your starting point? Write a formula that relates the radius and the circumference. Write a formula that relates the diameter and the circumference.

9 Cut a circle from heavy cardboard and place a string around its circumference. Measure the length of the string and the diameter of the circle. Is the circumference about $3\frac{1}{7}$ times the diameter? Is this answer consistent with the results of your other experiments?

10 Draw a figure similar to the spiral, but larger. Measure the length of the curve (a) with a ruler; (b) with a flexible tape measure; (c) by tracking the line with string and measuring the string; (d) by rolling a coin around the curve and counting the revolutions of the coin. (Find the circumference of the coin, too).

Experiment 5B(10)

11 Suggest and try another method for measuring this curve. Of all the methods tried, which was best in terms of accuracy? Worst? Which was easiest to do? Hardest?

12 Consider the angles shown. Which is smallest? Largest? Can you use a linear measure to rank them in order?

150

Experiment 5B(12)

13 Repeat Experiment 5B(9) for three more circular cutouts of various sizes. Are the ratios consistent? Further experimentation would show us that the ratio of the circumference to the diameter is not exactly $3\frac{1}{7}$ but rather is 3.141593 to seven significant digits. We represent this number by the symbol π. The circumference of a circle is found by multiplying the diameter by π.

14 Use a 1-foot ruler to measure the distance across a room. Repeat and compare your measurements.

15 Measure across a room using a yardstick or tape measure. Compare to Experiment 5B(14).

16 Measure the diagonals of a rectangular room both ways. Are they the same?

17 Devise your own unit of length and find the dimensions of this book using this unit of length.

18 Using your own foot, measure a distance on the floor. Repeat the measurement using a foot rule. How are the measurements related? Do you carry a good measuring device with you every place you go?

EXERCISE 5B

1 Relate the Archimedean principle to the situation of:
   (a) A man walking with 3-foot strides from New York to San Francisco
   (b) Measuring a room with a 1-foot ruler
   (c) A miler racing around an oval track four times
   (d) Removing the grains of sand from a beach
   (e) Counting the number of drops of water in the seas

2 Approximate the lengths of the following polygonal curves using four intermediate points, then eight and twelve intermediate points.

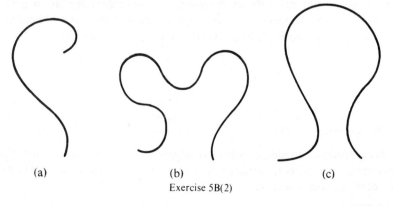

(a)　　　　　　　　　(b)　　　　　　　　　(c)

Exercise 5B(2)

151

3 Approximate the measures of the curves below using (a) a flexible measuring tool; (b) a 10-interval segment approximation.

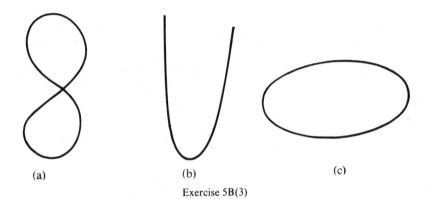

(a)   (b)   (c)

Exercise 5B(3)

4 Let C represent the circumference of a circle and d its diameter. Write a formula describing C in terms of d.

5 Let C represent the circumference of a circle and r its radius. Write a formula describing C in terms of r.

6 If a rectangle is a units long and b units high, what is a formula for its perimeter?

7 If a square has a side of s units, what is a formula for its perimeter?

8 Write a formula for the perimeter of a parallelogram with a base of b units and a height of h units, when the short diagonal is equal to b.

5.3 ANGULAR MEASURE

As in the development of linear measure, it is necessary for us to make some assumptions to develop the concept of angular measure. These are essentially modifications of the postulates about linear measure.

Postulate 5.5   There exists a correspondence between the points on a circle and the entire set of real numbers, that places the real numbers onto the circle.

Postulate 5.6   If any pair of arcs of a circle are congruent in one position, then they are congruent in any other position in which they may be placed.

We could use the idea of transformations to state Postulate 5.6—that is, any arc of a circle is invariant under the three transformations translation, rotation, and reflection.

152

Postulate 5.7   It is possible to subdivide an arc of a circle into as many subarcs as desired.

These three postulates enable us to talk about measures of angles. We have defined angles in terms of the union of rays, yet it is actually the opening *between* the rays to which we wish to assign a measure. We get around this difficulty by choosing a convenient radius $\overline{OA}$ (see Illus. 5.9) and constructing a circle with center at the vertex of $\angle AOB$. We associate this circle with the three postulates; since arc $\widehat{AB}$ of the circle lies in the interior of the angle, we can measure the angle by this arc. To make things simpler, we usually let the radius of the circle be 1; the circle is then called a *unit* circle.

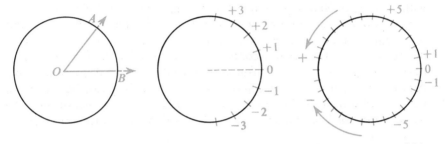

ILLUSTRATION   5.9
An angle *AOB* with vertex *O* at the center of a circle

ILLUSTRATION   5.10
A mapping of integers onto a circle

ILLUSTRATION   5.11
The directions of mapping for positive and negative numbers

Postulate 5.5 assumes a mapping of the real numbers onto (not a one-to-one correspondence, please note) the circle. This is done by letting the number 0 correspond to the intersection of the initial side of the angle and the circle (see Illus. 5.10) and assigning the positive numbers counterclockwise, and the negative numbers clockwise, from this point in such a way that the arcs between consecutive integers are congruent. Postulates 5.6 and 5.7 refer to these arcs.

The number placement and direction are arbitrary. The negative and positive numbers will overlap on the left side of the circle no matter how small we make the arc divisions (Illus. 5.11). Remember we did not assume a one-to-one correspondence between the points on the circle and the set of real numbers, but assumed we could match the real numbers with *some* points on the circle. Thus the overlap is no problem; it merely means that a given arc can have infinitely many measures. We need this concept of *retracing*, since an angle can be formed by rays that revolve around more than one complete circle.

153

Definition 5.2 Given any angle *AOC*, with real numbers *a* and *c* associated with the points *A* and *C* located on a unit circle with center at *O*, then $|c - a|$ is a measure of angle *AOC*.

We now need a basic unit (arbitrarily selected) with which to compare angles. One common measure is the angular *degree*. Another measure, the *radian*, is used frequently in theoretical mathematics. Other special applications require other basic units such as the *mil*.

The degree is an angle measure inherited from the ancient Babylonians who considered a year to have 360 days. They used this notion in segmenting a circle. If we partition the circumference of a circle into 360 congruent arcs, then the angle formed by two adjacent partitioning rays is an angle of the basic unit, 1 degree. A right angle thus is an angle with measure of 90 degrees and a straight angle measures 180 degrees.

We can measure all other angles by comparing them to this basic unit of 1 degree. Obviously the concept of congruence is important. One device for measuring angles contains a portion of a circular arc. This device, called a *protractor*, is usually 180 degrees or 360 degrees in scope. A protractor which contained just 1 degree could be used, but it would be very awkward to hold and would introduce errors in measuring. To use a protractor like that in Illus. 5.12, position the protractor over the

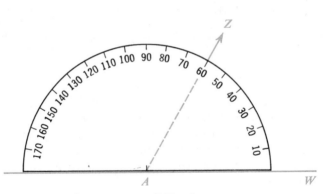

ILLUSTRATION 5.12   A protractor

angle to be measured so that the vertex of the angle is under the point marked *A* and one side of the angle is along the straight edge of the protractor. The measure of the angle can be read directly from the scale where the other edge of the angle intersects the protractor's curved surface. In Illus. 5.12, ∡*ZAW* measures 60 degrees.

154

If an angle is constructed in a circle (Illus. 5.13), such that the lengths of the radii $\overline{OA}$ and $\overline{OB}$ are equal to the length of the enclosed arc $\overarc{AB}$, this angle, $\measuredangle AOB$, is said to measure 1 radian, approximately equal to $57\frac{1}{3}$ degrees.

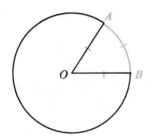

ILLUSTRATION 5.13   An angle of
measure 1 radian

ILLUSTRATION 5.14 $\measuredangle XYZ$,
$\measuredangle X'Y'Z'$, and $\measuredangle X''Y''Z''$ have
the same angle measure

Note that we have actually defined angular measures in terms of ratios of arcs to radii, so the radius doesn't matter. Consequently angles XYZ, $X'YZ'$, and $X'YZ'$ in Illus. 5.14 all have the same angular measure.

EXPERIMENT 5C

1   Suppose a circle was divided into ten equal parts around the circumference as shown here, and each arc called 1 *dec*.
   (a) Would it still be possible to measure angles accurately?
   (b) What would be the conversion of 1 dec to degrees? One degree to decs?

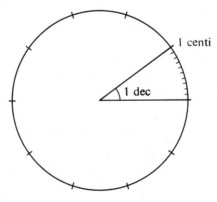

Experiment 5C(1)

155

2   Measure each of the angles shown using the dec scale.

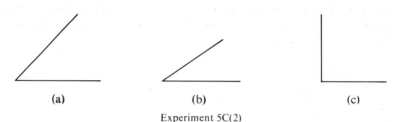

(a)   (b)   (c)

Experiment 5C(2)

3   Measure each of the angles in Experiment 5C(2) again using a standard pro-
    tractor. Convert your dec measurements to degrees and compare with the
    measurements you took in degrees.

4   The scale in Experiment 5C(1) also show 1 *centi*, where 1 dec = 10 centi.
    Compute the number of degrees in 1 centi. Measure each of the angles in
    Experiment 5C(2) in centis. Convert these measurements to degrees. Do you
    think it is better to measure in decs or centis?

5   The diagram illustrates a rather unorthodox protractor. Draw a facsimile
    of it to measure the angles in Experiment 5C(2). Compare your answers
    with the answers in Experiment 5C(2).

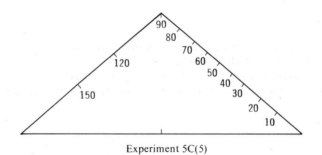

Experiment 5C(5)

6   Design a protractor that is rectangular in shape. Use it to measure the angles
    in Experiment 5C(2). Do your answers agree with those in Experiment 5C(2)?

7   Could the shape illustrated here be used for designing a protractor? What
    disadvantages would it have?

Experiment 5C(7)

8 In the illustration measure ∡AOB (in degrees); then measure ∡BOC. Measure ∡AOC. Compare the measure of this angle with that of each of the other angles and with the sum of the measures of the other angles.

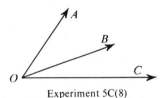

Experiment 5C(8)

9 In the illustrations for Experiment 5C(8), use $\overline{OA}$ as an initial line and measure ∡AOC. Then use $\overline{OC}$ as a base line and measure ∡AOC. How do your measurements compare? Draw another angle and repeat the procedure. Can you generalize?

10 Draw a circle and one radius labeled $\overline{OA}$. Draw a second radius $\overline{OB}$ so that $\overline{OA}$, $\overline{OB}$, and $\widehat{AB}$ are the same length. How did you determine the position of B? Did you use a flexible rule to help? Repeat this experiment for a circle of a different size. Label the points on it O', A', and B'. Is ∡$AOB \cong$ ∡$A'O'B'$?

EXERCISE 5C

1 Why might it be difficult for some children to decide that ∡A is larger than ∡B?

Exercise 5C(1)

2 Would it be better if the circle were divided into 100 parts or 1000 parts rather than 360? Can you name one advantage and one disadvantage of each of these division schemes?

3 Do you think it is possible to use *any* subdivision as the unit arc for measuring angles?

4 In the figure are two angles, ∡CBA and ∡$CB'A$. Which is larger? How do you know?

Exercise 5C(4)

5 What is the sum of the measures of the angles in a triangle in degrees? In radians?

157

6 Using your experiments in Chapter 3, what is the sum of the measures of the angles in a quadrilateral? A pentagon? A hexagon?

7 In the parallelogram what is $m(\angle ABC) + m(\angle BCD)$? Is this sum the same in any parallelogram?

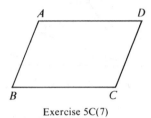

Exercise 5C(7)

8 Complete the following
(a) If $\angle A \cong \angle B$, then $m(\angle A)$____$m(\angle B)$.
(b) If $m(\angle A)$ is less than $m(\angle B)$, and $m(\angle B)$ is less than $m(\angle C)$, then $m(\angle A)$____$m(\angle C)$.
(c) If $\overrightarrow{OA}, \overrightarrow{OB}$, and $\overrightarrow{OC}$ are three rays, and $\overrightarrow{OB}$ is between $\overrightarrow{OA}$ and $\overrightarrow{OC}$, then $m(\angle AOB) + m(\angle BOC) = m(\_\_\_)$.

9 Is $m(\angle AOB) = m(\angle BOA)$?

10 Measure the angles in the figure using a protractor. Compare the measures of the angles of the two smaller triangles. Are the triangles similar?

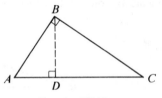

Exercise 5C(10)

11 Fill in as many angle measures as you can in each of the figures, using properties developed in earlier chapters. Check your answers by actually measuring the angles.

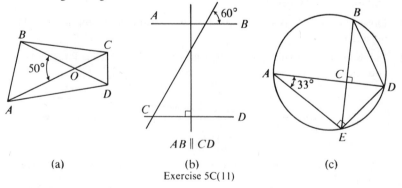

(a) (b) (c)
$AB \parallel CD$
Exercise 5C(11)

158

12 Construct regular polygons of three, four, five, six, seven, eight, nine, and ten sides. Calculate the size of the central angle determined by one side in each.

5.4   AREA MEASURE

The concept of measurement on a surface is certainly a familiar one, but it contains problems not encountered in linear or angular measure. It is reasonable for a man to survey his lands and wonder, "How much land do I have?," but if he compares different portions of his land, asking, "Which section is the largest?," he is posing a question that might be difficult to answer. In Illus. 5.15 are three differently shaped plots all of which have the same measure but which do not look alike.

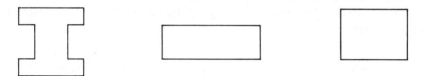

ILLUSTRATION 5.15   Three plane figures with the same area

We need only look at a map of the United States to visualize the problem of deciding which of several states is the largest. The nonsimilar shapes of the various states makes it difficult to rank them in size.

To develop the concept of measure on a surface, which we call *area*, and to be consistent in our ideas on measure, it is desirable to relate surface measure to linear measure.

Definition 5.3   Given any portion of a plane, we can associate with it a real number called its measure or *area*. This area is given in square units.

Any portion of a plane is called a *region*. We need some basic arbitrary unit with which to compare regions and to use to measure the area of a region. We choose a square unit of arbitrary size and assign it an area of 1. The basic unit need not be square but could be triangular or hexagonal; however, the area in Illus. 5.16 can be measured more easily with square units than with any other.

In Illus. 5.16(a), a length of 7 units and a width of 5 units yields an area of $7 \times 5 = 35$ square units (count them). In (b) a length of $3\frac{1}{2}$ rectangular units and a width of 5 units has an area of $3\frac{1}{2} \times 5 = 17\frac{1}{2}$ rectangular units. In (c), we cannot say that a units of length and b units of width give ab units of area; we would have to count the triangular units to find the area. We maintain consistency with linear measure by

159

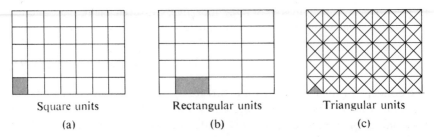

Square units   Rectangular units   Triangular units

(a)   (b)   (c)

ILLUSTRATION 5.16   Possible measuring units in the plane

using the linear unit as the length of the side of each square. Although there is a relationship between the two measurements, *one unit of length is an entirely different kind of unit from one unit of area*.

We can formalize our idea of area of rectangular regions by stating the following:

Formula 5.1   The total number of units of area of a rectangular region *a* units long and *b* units wide is *ab* square units. Thus $A = lw$, where A is the area of the region, l is its length, and w is its width.

From this one formula we can easily derive formulas for the areas of other regions. A figure closely related to the rectangle is the parallelogram. In Illus. 5.17, we can see that a parallelogram with base of 7 units and height of 5 units has exactly the same area as a rectangle of the same dimensions. (Move the shaded triangle from right to left to form a rectangle.) Later we shall examine other parallelograms; in each case the relation will be the same. Notice the height is the *perpendicular* height, not the distance measured along an edge.

Formula 5.2   The total number of units of area of a parallelogram with base *b* units long and height of *h* units is *bh* square units.

Illustration 5.18(a) is a right triangle that has a base of 6 units and a height of 4 units. Superimposed on it is a rectangle of dimensions 6 and 4.

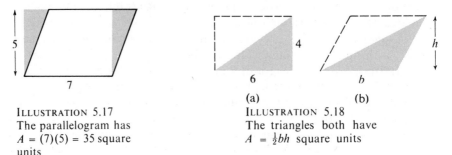

(a)   (b)

ILLUSTRATION 5.17
The parallelogram has
$A = (7)(5) = 35$ square
units

ILLUSTRATION 5.18
The triangles both have
$A = \frac{1}{2}bh$ square units

160

By inspection, the area of the triangle is exactly one-half the area of the rectangle. It is true for any right triangle, that is, the area of the triangle is equal to one-half the area of the rectangle with the same base and height. Illustration 5.18(b) shows a general triangle on which a parallelogram has been superimposed. Inspection reveals the area of the triangle to be one-half the area of the parallelogram. Since it is always possible to superimpose a parallelogram on a triangle as in Illus. 5.18(b), we can generalize our results.

Formula 5.3 The total number of units of area of any triangle with base of b units and height of h units is $\frac{1}{2}bh$ square units.

Illustration 5.19 shows three triangles, each with a height of h units and a base of b units. Although these triangles are quite different in appearance, they all have the same area: $\frac{1}{2}bh$.

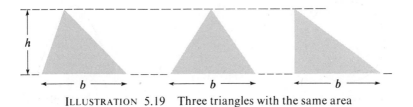

ILLUSTRATION 5.19 Three triangles with the same area

We can extend our discussion to find a formula for the area of a trapezoid. Illustration 5.20 shows a trapezoid in solid lines and a second trapezoid, adjacent and congruent to the first, in dashed lines. In this

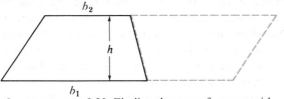

ILLUSTRATION 5.20 Finding the area of a trapezoid

juxtaposition they form a parallelogram. The base of this parallelogram is congruent to the sum of the two bases of the original trapezoid, and the altitude of the parallelogram is the altitude of the trapezoid. The formula for the area of any parallelogram is base × height; hence the area of this parallelogram is $h(b_1 + b_2)$. Since the area of the trapezoid is exactly one-half the area of the parallelogram, it has an area of $\frac{1}{2}h(b_1 + b_2)$.

161

Although we were referring to the specific trapezoid in Illus. 5.20 as we made this development, the notation we used was general, and consequently our statements apply to any trapezoid. This allows us to formalize another statement about area.

Formula 5.4 The total number of units of area of a trapezoid of height h with bases of lengths b_1 and b_2 is one-half the height times the sum of lengths of the bases; that is, $A = \frac{1}{2}h(b_1 + b_2)$.

The development of other area formulas is left for the exercises.

You will recall that we encountered some difficulty in measuring curved lengths with linear measures. Similarly, difficulties arise in measuring regions bounded by curves with square units—the areas of the geometric figures in Illus. 5.21 are difficult to obtain using square units.

ILLUSTRATION 5.21 Irregularly shaped geometric figures in the plane

It should be obvious that no combination of squares will be congruent to these figures. However, experience tells us that square units are still the best units for area measure.

Consider the circle. We can approximate the area of a circle in square units by counting squares in the interior of the circle. We see in Illus. 5.22 that there are 32 squares totally within the boundary. Certainly the area of the circle is more than this, since there are portions of other squares which were not counted. We can say that the area of the circle is greater than 32 square units.

A further inspection reveals that the region of the circle is contained within the region designated in Illus. 5.22 by the shaded squares. A count of these yields 52. We can now say that the area of the circle is less than 52 square units, the area of the shaded region. We have an approximation of the circular area since $32 < A < 52$. This is not very accurate. A better approximation can be found by making the squares smaller and increasing their number. In Illus. 5.23, each square is one-quarter the area of an original square. By counting we find that there are 148 of these squares inside the circle, and it takes 188 to contain the circle. Since four of these squares yield 1 square unit, the area of the circle must be between $\frac{148}{4}$ and

162

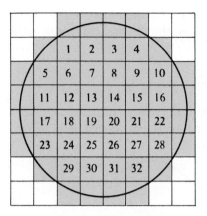

ILLUSTRATION 5.22 Approximating the area of a circle

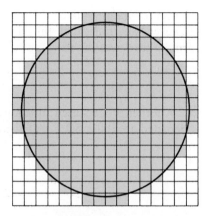

ILLUSTRATION 5.23 A better approximation to the area of the circle

$\frac{188}{4}$, or $37 < A < 47$. Further refinements can be made so that the area can be determined to any degree of accuracy in terms of square units. However, an area found in this way is always approximate.

We can derive a formula for the area of a circle by using a somewhat different approach. Consider the circle in Illus. 5.24(a); it has been partitioned into 12 equal sections. These sections have been reassembled in Illus. 5.24(b) to make a figure that is approximately the shape of a parallelogram. If the wedges are made much smaller, the ends of the figure formed when the pieces are reassembled will be very nearly vertical and the configuration very closely resembles a rectangle. Since the circumference of the circle as determined in Experiment 5B is $2\pi r$, the base of the rectangle is πr. Do you see why? The rectangle also has a height of r. Using the formula for the area of a rectangle we find $A = bh = (\pi r)r = \pi r^2$. The proof that this is indeed the correct formula for

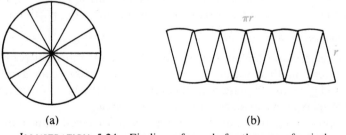

(a)   (b)

ILLUSTRATION 5.24   Finding a formula for the area of a circle

163

the area of a circle involves techniques beyond the scope of this book. However, we think that this development should make the following formula seem reasonable.

Formula 5.5 The total number of square units of area A of a circle with radius of measure r is $A = \pi r^2$.

Another process used for approximating the area of irregular shapes is depicted in Illus. 5.25. The given figure is partitioned into several

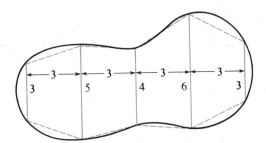

ILLUSTRATION 5.25   Approximating area by trapezoidal partitions

parts by parallel lines a uniform distance apart. The curved portions of the boundaries are then approximated by line segments. If small segments are used they should not differ from the true boundaries by much. In effect, the irregular region is then approximated by a series of trapezoids. Since the area of a trapezoid can be computed exactly in terms of square units, we can compute the approximate area of the region. Using the dimensions in Illus. 5.25, the approximation to the area of the region when four partitions are made is

$$\tfrac{3}{2}(3 + 5) + \tfrac{3}{2}(4 + 5) + \tfrac{3}{2}(4 + 6) + \tfrac{3}{2}(3 + 6)$$

$$= \tfrac{3}{2}(8) + \tfrac{3}{2}(9) + \tfrac{3}{2}(10) + \tfrac{3}{2}(9) = 54 \text{ square units}$$

This is not a very good approximation, for the line segments differ from the real boundaries rather noticeably in some sections. However, it would be easy to obtain a better approximation by making more partitions.

The Pythagorean Relationship

A basic relationship among the sides of a right triangle was observed by the Greek mathematician Pythagoras nearly 2500 years ago. It is known today as the Pythagorean theorem and is the basis for many geometric properties, particularly those involving area.

Suppose a square, 1 unit on a side, is constructed and a diagonal is drawn. What is the length of the diagonal (Illus. 5.26(a))?

164

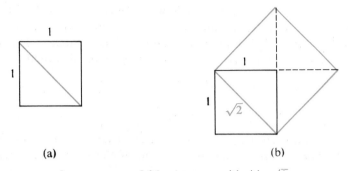

(a)　　　　　　　　　　　　　(b)

ILLUSTRATION 5.26　A square with side $\sqrt{2}$

If the diagonal is made the side of a second square, as in Illus. 5.26(b), then the area of the new square is 2 square units, since it contains four triangles congruent to the two triangles in the original. Since the area of a square is the square of a side, it follows that the side S when squared equals 2; that is $S^2 = 2$ or $S = \sqrt{2}$. Hence the diagonal must be $\sqrt{2}$ units long. This suggests the relationship in the original Illus. 5.26(a) that

$$1^2 + 1^2 = (\sqrt{2})^2$$

In a square 7 units on a side, lines can be drawn to form four right triangles as shown in Illus 5.27 so that the two shorter sides are 3 and 4 units long. All four triangles are congruent, and each has an area of $\frac{1}{2}(3 \cdot 4) = 6$ square units. This yields 24 square units in the triangles; since there are 49 square units in the large 7-by-7 square, there are $49 - 24 = 25$ square units in the interior quadrilaterial. It has right angles at each corner (do you see why?), hence it is a square. This means each side must be $\sqrt{25} = 5$ units long. It is no coincidence that $3^2 + 4^2 = 5^2$, where 3, 4, and 5 are the lengths of the sides of each triangle.

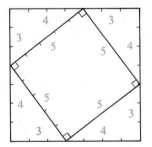

ILLUSTRATION 5.27　The Pythagorean relationship for $3^2 + 4^2 = 5^2$

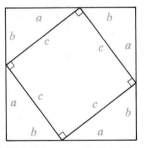

ILLUSTRATION 5.28　The general Pythagorean relationship

165

Consider now the general case shown in Illus. 5.28. Each of the four congruent right triangles has an area of $\frac{1}{2}ab$, giving a total of $2ab$ square units in the triangles. The interior square contains c^2 units. The overall square has an area of $(a + b)^2 = c^2 + 2ab$. Simplified, this yields

$$a^2 + 2ab + b^2 = c^2 + 2ab$$

or

$$a^2 + b^2 = c^2$$

where a, b, and c are the sides of any right triangle with c as the hypotenuse. This completes our justification of the Pythagorean relationship.

Formula 5.6 In any right triangle with sides of lengths a and b, and where c is the hypotenuse, $a^2 + b^2 = c^2$.

EXPERIMENT 5D

1   Add the area contained entirely within the circle and the area containing the circle in Illus. 5.22 and divide by 2. Does this suggest a way to approximate the area of a curved region? Why does this method work?

2   Compare the areas in the figures. What conclusions can you draw about shape versus area?

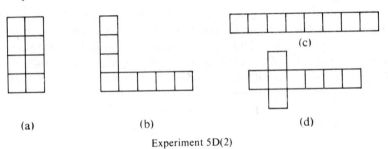

(a)   (b)   (d)

Experiment 5D(2)

3   Make a model of the triangle illustrated in (a) and use it to measure the figures in (b) and (c). Can you see any advantage in using a triangle as the basic unit for area measure?

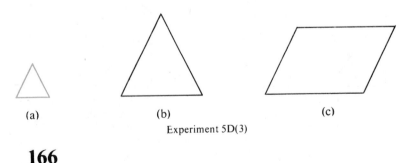

(a)   (b)   (c)

Experiment 5D(3)

166

4 Try to measure the figures in Experiments 5D(3b) and (3c) using a model of the circle as a basic unit. What is the problem with using a circle for a basic unit?

Experiment 5D(4)

5 Use the parallelogram to measure the figures in Experiments 5D(3b) and (3c). Can you tell when this kind of unit might be convenient?

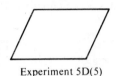

Experiment 5D(5)

6 Make a cutout of the parallelogram, cut along the dotted line, and fit the cut piece to the opposite side. What figure is produced?

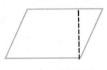

Experiment 5D(6)

7 Repeat Experiment 5D(6) for several parallelograms. Make sure the dotted line is perpendicular to the side. Does the cut piece always "fit" the other side? What is the name of the new figure? Does this strengthen an earlier result? Prove deductively that the reassembled parallelogram is always a rectangle.

8 Draw several parallelograms and one diagonal of each. Cut along the diagonal and superimpose the two resultant triangles. Does your observation make you believe the area of a triangle is always $\frac{1}{2}bh$?

9 Given a regular hexagon in which the length of one side is 5 units, can you find its area using regular triangles?

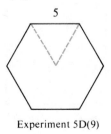

Experiment 5D(9)

167

10 Can you suggest a general formula for the area of a regular hexagon?

11 On a large sheet of paper or on a chalkboard, construct a circle with a radius of 10 inches.
   (a) Approximate the area of one-quarter of the circle with a triangle as shown. Measure carefully.
   (b) Approximate the area of one-quarter of the circle with a rectangle and a triangle as shown. Measure carefully.
   (c) Repeat the experiment with five partitions as shown. Measure carefully.
   (d) Repeat the experiment with ten partitions. Measure carefully.
   (e) Compare the four answers.
   (f) Multiply each by 4 and get successive approximations to the area of the entire circle.
   (g) How does each of the answers compare to 314 square inches?

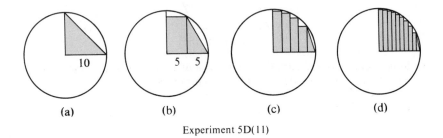

(a) (b) (c) (d)

Experiment 5D(11)

12 Repeat Experiment 5D(11) using a circle of radius 8 inches. Compare your answers to an area of 201 square inches. (Use eight partitions instead of ten).

13 Repeat Experiment 5D(11) using a radius of 5 inches. Compare your answer to an area of 78 square inches.

14 Do the preceding experiments suggest a ratio between the area and radius of a circle?

15 Make a copy of Illus. 5.28 on paper. Cut the triangles and inner square from the paper, and see if you can reassemble them as shown. Show how this also establishes $a^2 + b^2 = c^2$.

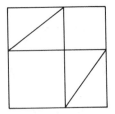

Experiment 5D(15)

EXERCISE 5D

1   Why will the area of the region determined by the numbered squares in Illus. 5.22 always be less than the true area? Is the second method for finding circular area suggested in the experiments better than the first?

2   Would an oval shape be a good shape to use for a basic unit for measuring area? Why or why not?

3   Use a basic unit of $\frac{1}{5}$ inch to compute the areas of the figures.

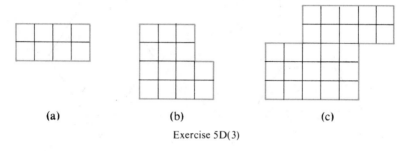

(a)   (b)   (c)

Exercise 5D(3)

4   Use the approximation processes described in this section to find the areas of the regions enclosed by the curves, using squares $\frac{1}{4}$ inch on a side.

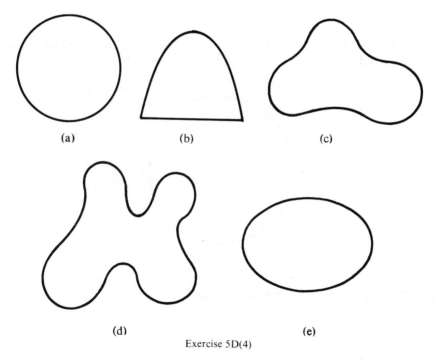

(a)   (b)   (c)

(d)   (e)

Exercise 5D(4)

169

5   Partition each of the figures for Exercise 5D(4) into four regions that might be used to find the area of each figure. Then subdivide each region to form eight regions. Which division is better? What difficulties do you encounter?

6   Find the areas of the figures using formulas.

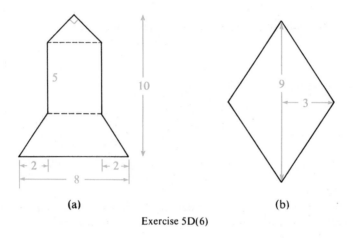

(a)   (b)

Exercise 5D(6)

7   Approximate the area of the figure according to the five partitions (measure the chords).

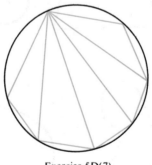

Exercise 5D(7)

8   Find the missing lengths for each triangle.

170

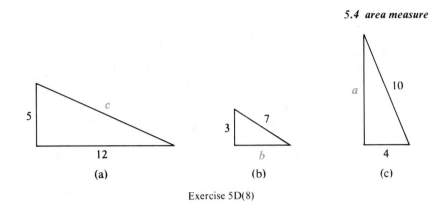

(a)   (b)   (c)

Exercise 5D(8)

9 Calculate the height of the equilateral triangle in the circle with radius 10. Use the Pythagorean relationship.

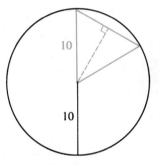

Exercise 5D(19)

10 Find *m* and *n* in the illustration.

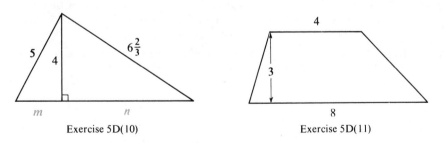

Exercise 5D(10)   Exercise 5D(11)

11 Find the area of the trapezoid.

171

12 What happens to the area of a rectangle if the length is doubled but the width is not changed? If the length is tripled and the width doubled? If the length is made one-half the original and the width doubled?

13 If the area of a circle is 7.854 square units, what is the radius?

14 Draw four differently dimensioned rectangles, each with an area of 24 square units.

15 Each of the equal sides of an isosceles triangle is 10 inches, and the base is 16 inches. What is its perimeter? Its area?

16 If the radius of a circle is doubled, how is the circumference changed? The area?

17 If each side of an equilateral triangle is doubled in length, how is the perimeter changed? The area?

18 Find the area of the shaded region.

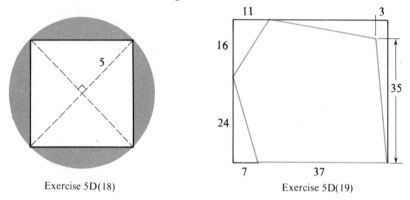

Exercise 5D(18)   Exercise 5D(19)

19 Find the area of the polygon inside the rectangle.

20 A rectangle has a perimeter of 16 inches. Give the dimensions of several rectangles which have such a perimeter. Which has the largest area? What special case of a rectangle produces maximum area?

21 Repeat Exercise 5D(4) using trapezoids with height $\frac{1}{4}$ inch. Which process gives the better approximation?

172

6 / MEASURE OF VOLUME AND SURFACES OF SOLIDS

6.1 VOLUME AND CUBIC MEASURE

Three-dimensional measure is a further extension of one-dimensional measure. Thus, the measure of a one-dimensional object (a line or curve) is called length; the measure of a two-dimensional object (a plane region) is called area; and we now define the measure of a three-dimensional object to be *volume*. That is, we can assign to a given region of space a measure called its *volume*. Remember that it is not the region that has a numerical value, but rather the *volume* of the region.

When we measured area of figures in a plane, we introduced a new measure concept, the square unit. The extension of measure to solid regions in space also requires a new concept, the *cubic* unit.

The basic unit of measure of the solid region, the cubic unit, is the extension of the linear unit into three-dimensional space. The volume of a cube with measure 1 linear unit on each side is defined to be 1 cubic unit; all measures of space are made in terms of this basic unit. The cubic unit fills two needs of volume measure: it is "space filling," and it is easy to use in calculations. A stack of blocks in a box gives intuitive evidence of the first need. A box filled with balls, for example, has many open spaces between the balls; hence, all the space in the box is not filled.

Ease of calculation of volume is a result of the cube having square faces, each face at right angles to the adjacent ones. This means the cube has a linear measure in each of the three dimensions, at right angles to each other. This in turn yields a quick means of calculating volumes

173

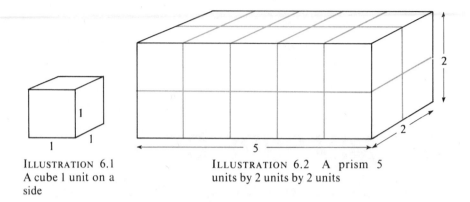

ILLUSTRATION 6.1
A cube 1 unit on a
side

ILLUSTRATION 6.2 A prism 5
units by 2 units by 2 units

of figures bounded by polygons, especially polygonal faces at right angles to each other. Thus, the total number of units of volume in the solid in Illus. 6.2, which is 5 units long, 2 units high, and 2 units wide, can be found by partitioning it into unit cubes and counting the cubes. The total is 20. This same value can be obtained by observing that there are 2 layers in the solid, and that each layer has 2 rows of 5 cubes each. The product $2 \times 2 \times 5 = 20$ is then the number of units of volume in this solid. (In the formulas and discussion to follow, we shall write "volume" rather than "number of units of volume," where the meaning is understood.)

Formula 6.1 The volume V of a rectangular solid (right rectangular prism) is given by $V = lwh$, where l, w, and h are the length, width, and height, respectively.

Finding the volumes of other solids, such as those pictured in Illus. 6.3, poses a more difficult problem, but it can be done by using an approach similar to that used in approximating the areas of irregular plane figures. We use the cubic unit as the basis of comparison for all these

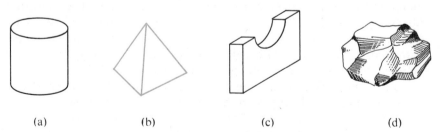

(a)   (b)   (c)   (d)

ILLUSTRATION 6.3   Irregular three-dimensional figures

174

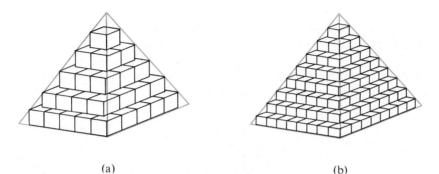

(a)   (b)

ILLUSTRATION 6.4   Approximating volume by cubes

figures. For example, a solid such as a pyramid can be approximated by
a set of cubes as in Illus. 6.4(a). The volumes of the pyramid and the stack
of cubes are obviously not the same; closer agreement between the volumes is obtained by decreasing the size of the cubes and increasing the
number of cubes as in Illus. 6.4(b).

The same process could be used for finding the approximate volumes
of the other figures in Illus. 6.3. In general, it is not possible to find the
exact volumes of these figures (those with curved faces) by filling them
with cubes (which have plane faces), but we can approximate them closely.
We caution that this method is laborious and should only be used when
other methods fail.

Formulas for the volumes of some general polyhedra can be found by
relating them to figures with right angles. Illustration 6.5(a) shows a rectangular prism with lateral faces not at right angles to each other. It is
not possible to find the volume of this figure by filling it with unit cubes,
since its faces would not be parallel to the faces of the cubes. However,
if we think of this figure as being made up of many sheets of paper (Illus.
6.5(b)), its volume is the sum of the volumes of all the sheets of paper.
Moreover, the paper can be pushed into a right rectangular stack as in

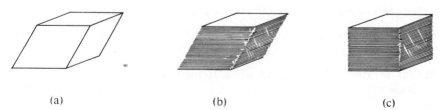

(a)   (b)   (c)

ILLUSTRATION 6.5   Finding the volume of a prism with parallelograms for faces

175

Illus. 6.5(c), without loss or addition of volume. Since the new figure is a right rectangular solid it has volume equal to length times width times height, and therefore so does the original prism.

It is also possible to partition a prism at right angles to the base to establish a formula for volume. The prism in Illus. 6.6(a) is partitioned in (b), and in (c) the pieces are moved to form a right rectangular solid.

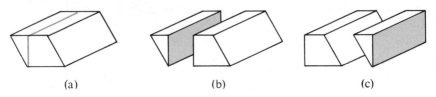

(a)   (b)   (c)

ILLUSTRATION 6.6   Partitioning a parallelepiped

As long as opposite faces are parallel, such a manipulation can be made, so the volume of any quadrilateral prism with opposite faces parallel can be calculated.

Formula 6.2   The volume V of any rectangular prism is given by $V = lwh$, where l and w are the length and width and h is the height perpendicular to the base.

Notice that h is not necessarily the measure of a lateral face. As a corollary to this formula we state

Formula 6.3   The volume V of any parallelepiped (opposite pairs of faces parallel) is given by $V = lwh$, where only one of the measures l, w, or h is necessarily a dimension of a face.

Another result of Formula 6.2 can be derived from the fact that the volume of a rectangular prism can be considered as

$$V = (lw)h = A_b h$$

where A_b is the area of the base. Then Formula 6.2 can be extended to include all prisms, whether their bases are rectangular or not:

Formula 6.4   The volume V of any prism is $V = A_b h$, where A_b is the area of the base and h is the measure of the altitude.

In Illus. 6.7 the volume of either of the triangular prisms can be calculated as follows:

$$V = A_b h \quad \text{or} \quad V = [\tfrac{1}{2}(4)(5)](6) = 60 \text{ cubic units}$$

176

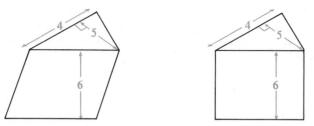

ILLUSTRATION 6.7   Two triangular prisms with the same volume

Note that the height of the prism may not be an edge of the prism. Formulas for related solids are developed in the following experiments and exercises.

EXPERIMENT 6A

1   Partition the solid to make two solids, and put them together to form a right rectangular solid. Can you now write the formula for the volume of each of the solids? Since you could consider the base of the original solid to be a triangle, do you think there is verification for your formula?

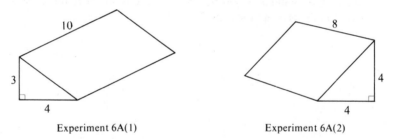

Experiment 6A(1)   Experiment 6A(2)

2   Can you suggest a method other than Formula 6.4 for finding the volume of the prism shown above by partitioning it to make a rectangular solid? Assume it is a right prism.

3   Make four "triangular prisms" using the pattern, and fit them together to form a cube as shown. Can you write the formula for the volume of each of these solids?

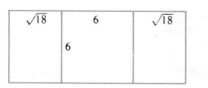

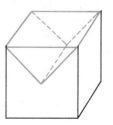

Experiment 6A(3)

177

4 Suppose you had six of the pyramids pictured so that they would fit together to form a cube in which the base of each pyramid is a face of the cube. Can you find the volume of the pyramid? Does this suggest a possible formula for the volume of such a pyramid?

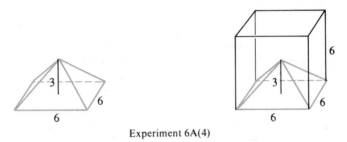

Experiment 6A(4)

5 Given two containers, one in the shape of the pyramid suggested in Experiment 6A(4) and another in the shape of the cube, suggest a way to verify empirically that the volume of the pyramid is one-sixth that of the cube.

6 Given a pyramid and a prism with right triangular base as shown, what do you think the volumes might be? Suggest a way to verify this experimentally with containers in these shapes.

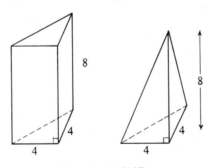

Experiment 6A(6)

7 Given a lump of coal or a rock shaped somewhat like the illustration, what ways could you suggest to find its volume?

Experiment 6A(7)

178

8 Consider a triangular prism with the given dimensions. A cut along the line 2 units above the base and parallel to the base results in a partition that can give a rectangular solid. Do you see how? What is the volume of the rectangular solid and hence of the triangular prism? Can you suggest a formula for the volume of a triangular prism?

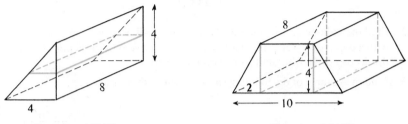

Experiment 6A(8)   Experiment 6A(9)

9 Can you suggest a formula for the volume of this trapezoidal prism?

10 Can you generalize your results in Experiment 6A(9) for any isosceles trapezoidal prism?

11 Can you suggest a partition of the figure that would yield its volume? Another partition? What is its volume?

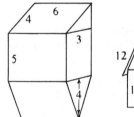

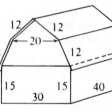

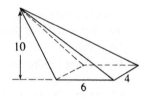

Experiment 6A(11)   Experiment 6A(12)   Experiment 6A(14)

12 A familiar barn shape is shown. Suggest a means for partitioning it to find its volume; then calculate the volume.

13 Demonstrate, using styrofoam or clay, that the figure in Illus. 6.6(a) can be reassembled as in (c).

14 Suggest three ways to approximate the volume of the oblique pyramid shown.

EXERCISE 6A

1 Calculate the volume of each of the solids, partitioning if necessary. Suppose the linear units were changed to yards from, say, feet or inches. Would the numerical answers be the same?

179

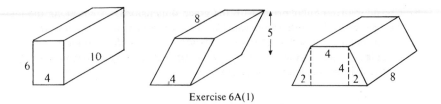

Exercise 6A(1)

2 Find the volume of a rectangular swimming pool 50 feet wide, 120 feet long, with a depth of 3 feet at the shallow end sloping evenly to a depth of 11 feet at the deep end. In what units is its volume given?

3 Find the volume of the building shown. Dimensions are in feet.

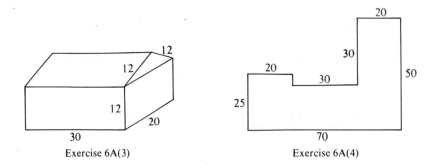

Exercise 6A(3)   Exercise 6A(4)

4 Calculate the number of cubic *yards* of dirt to be removed to dig a basement for a house if it is in the shape shown and is to be 10 feet deep. All dimensions are in feet.

5 What is the number of cubic inches of volume of a deep-freeze unit whose interior dimensions are 22 inches by 30 inches by 54 inches? Restate your answer in cubic feet (1728 cubic inches = 1 cubic foot).

6 In Exercise 6A(2) calculate the number of gallons of water needed to fill the pool full if $7\frac{1}{2}$ gallons = 1 cubic foot.

7 Derive a formula for finding the volume of a regular pentagonal prism.

8 Derive a formula for finding the volume of a regular hexagonal prism.

9 A house has 1300 square feet of floor space and ceilings 8 feet 3 inches high. What is the volume of the house?

10 In Exercise 6A(9), the house has an attic fan that moves 400 cubic feet of air per minute. How long would it take to change the entire volume of air in the house?

11 If the steps shown on the next page are to be 3 feet wide, calculate the number of cubic yards of concrete needed to fill the shaded portion of the stairway form.

180

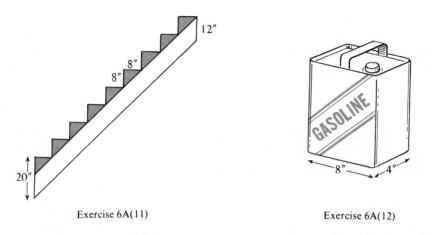

Exercise 6A(11) Exercise 6A(12)

12 A gallon is about 231 cubic inches. Using this conversion factor calculate the height of the rectangular 1-gallon container.

13 If the can in Exercise 6A(12) is to contain 2 gallons and the dimensions of the new can are in the same ratio as those in the 1-gallon can, what are its dimensions?

14 Calculate the volumes of these figures.

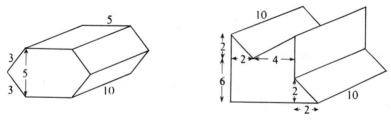

Exercise 6A(14)

15 Explain how Formula 6.3 follows from Formula 6.2.

6.2 CYLINDERS AND SPHERES

We face the same kind of difficulty in finding the volume of a cylinder as we did in finding the area of a circle. Area is measured in square units, but a circle has no linear boundary. Likewise, volume is measured in cubic units but a cylinder has no plane lateral faces. We can find the approximate volume of a cylinder by filling it with unit cubes. However, a simpler procedure suggests itself if we relate the cylinder and the prism: The definitions of these two solids are alike except for the generating curve; therefore we might surmise that their formulas for volume are alike

181

except for the method of finding the areas of the bases. The cross-sectional area of any "slice" through a cylinder is the same as long as the cross sections are parallel to the base. This uniformity of the cylinder yields the following formula:

Formula 6.5   The volume V of a cylinder is $V = A_b h$, where A_b is the area of the base and h is its height.

For a right circular cylinder, $A_b = \pi r^2$, and h is its height. This is suggested in Illus. 6.8(a). However, Formula 6.5 applies to all cylinders regardless of the angles of their major axes (as shown in (b) and (c)).

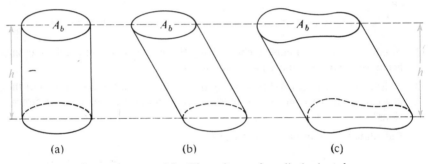

(a)   (b)   (c)

ILLUSTRATION 6.8   The volume of a cylinder is $A_b h$

We can generalize this concept to all solids with uniform cross-sectional area. The volume of any solid is $V = A_b h$ if its cross-sectional area, taken at any plane parallel to the plane of the base, is equal to the area of the base. It should be clear that cylinders and prisms fall into this category, as well as composite solids such as the one in Illus. 6.9.

Spheres, ellipsoids, and cones do not have uniform cross sections in any series of parallel planes, and their volumes cannot be evaluated by this formula. To calculate the volume of a sphere requires some ingenuity.

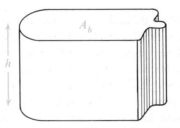

ILLUSTRATION 6.9   The volume of the composite is $A_b h$

182

We could approximate the volume of a sphere by considering it a poly-hedron of many faces, but calculating the volume of, say, a 20-faced regu-lar polyhedron is no simple task, and the result would only be an approxi-mation.

Another way to find the approximate volume of a sphere is to con-sider the sphere as a stack of cylinders or discs, each relatively "thin" compared to its diameter. The volume of each disc shown in Illus. 6.10 could be calculated, using the formula for the cylinder. Then, the volume

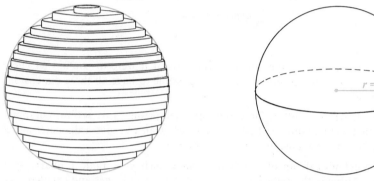

ILLUSTRATION 6.10 Approxi-mating the volume of a sphere by cylindrical layers

ILLUSTRATION 6.11 The volume of a sphere is $\frac{4}{3}\pi r^3$

of the sphere is approximately equal to the sum of the volumes of the discs. The greater the number of discs used, the better the approximation. Again, calculation of the separate volumes becomes tedious, and we will simply state here that the volume of a sphere can be approximated by such a method and the exact volume can be found by a limiting process. The interested student may consult a calculus text for a precise evaluation. The formula derived for the volume of a sphere is:

Formula 6.6 The volume V of a sphere with a radius of measure r is $V = \frac{4}{3}\pi r^3$.

Thus, if a sphere has a radius of measure 4 inches (approximately the size of a basketball), then its volume is $V = \frac{4}{3}\pi(4)^3 = \frac{256}{3}\pi$ cubic inches $\doteq$ 268.1 cubic inches (Illus. 6.11).

EXPERIMENT 6B

1 Suppose you wanted a cylinder and a square prism of equal heights to have the same volume. How could you relate their base dimensions? Can you

183

suggest an empirical way to determine the method for relating their base dimensions?

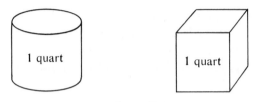

Experiment 6B(1)

2   What dimensions could a square prism have if it were to contain 64 cubic inches? List several combinations. Is it possible to have all three dimensions the same?

3   Repeat Experiment 6B(2) for a right circular prism. List several combinations that yield a volume of 64 cubic inches. Can you give a diameter and a height that are the same that yield this volume?

4   Find a cylindrical can of a stated volume, such as 1 pint, 1 quart, or 1 gallon. Measure the radius and height of the can in inches and calculate the volume of the can in cubic inches. Compare this to the designated volume of 1 gallon $\doteq$ 231 cubic inches (1 quart $\doteq$ 58 cubic inches, 1 pint $\doteq$ 29 cubic inches).

5   Compare two cylindrical cans of the same height with diameters in the ratio of 1:2. What is the relationship of their volumes?

6   Using the results of Experiment 6B(5), what conclusion could you draw about the volumes of cylinders with the same height and different diameters? Compare cylinders with diameters of 1, 2, 3, 4, and 6 if necessary.

7   Compare the volume of a cube with a 10-inch edge and a sphere with a 10-inch diameter. Does your result seem reasonable?

8   Compare the volume of a cylinder with altitude and diameter 10 inches to a sphere with $d = 10$ inches. Which is larger? What do the results of Experiments 6B(7) and 6B(8) tell you with reference to fitting given objects of these shapes inside one another?

9   Calculate and compare the volumes of spheres with radii of 5 inches, 10 inches, and 20 inches. Can you make a conclusion about the relationship of the volumes of two spheres if you know their radii?

10   Consider the sphere in the diagram. You could imagine the volume divided into many congruent solids like the one shown. If these solids have very small bases, approximately what is the height of each? Does this suggest another way for approximating the volume of a sphere?

184

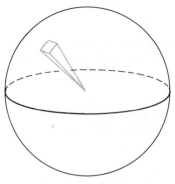

Experiment 6B(10)

11 Using a graduated container such as a laboratory cylinder or a measuring cup, find a way to determine the volume of an irregular object such as a stone.

EXERCISE 6B

1 Calculate the volumes of the cylinders shown. All are circular cylinders.

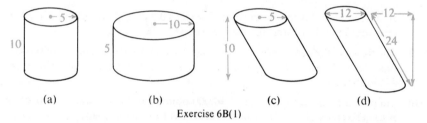

(a)   (b)   (c)   (d)

Exercise 6B(1)

2 Find the volumes of the solids shown. All prisms are rectangular; all cylinders are circular.

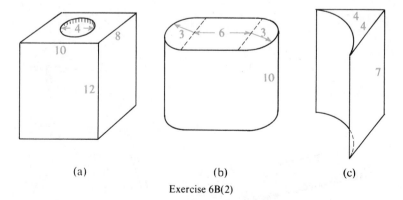

(a)   (b)   (c)

Exercise 6B(2)

3 Calculate the volume of a cylinder whose perimeter is 40 feet and is 12 feet high. Give your answer in cubic feet.

185

4 In Exercise 6B(3), if 1 cubic foot = $7\frac{1}{2}$ gallons, calculate the volume in gallons.

5 Using the information in Exercise 6B(4), how high would a cylindrical tank have to be to hold 10,000 gallons of water, if it is to have a 5-foot radius?

6 How many gallons would a pipe 6 inches in inside diameter and 1000 feet long hold?

7 The figure represents the cross section of a metal casting 12 inches long. The center hole is 1 inch in diameter. Calculate the volume of the casting in cubic inches.

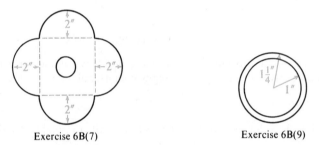

Exercise 6B(7) Exercise 6B(9)

8 In Exercise 6B(7), suppose the casting weighs 0.5 pound per cubic inch. What is its weight to the nearest tenth of a pound?

9 The figure is the cross section of a pipe. Calculate the volume of 300 feet of pipe. If the weight of the pipe is 0.2 pounds per cubic inch, what is the weight of the 300 feet?

10 Suppose a truck can carry up to 10,000 pounds and is to haul a load of the pipe in Exercise 6B(9). How many feet can it haul in one load?

11 Find the volume of the tank shaped as a cylinder with a hemisphere at each end, where the overall length is 20 feet and the height is 8 feet.

Exercise 6B(11)

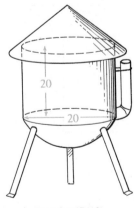

Exercise 6B(12)

186

12 A railroad water tower has a tank shaped as a cylinder with a hemispherical bottom as shown. Calculate the volume of the tank in cubic feet and gallons.

13 Find the volume of a sphere with a circumference of 30 inches; 50 inches; 100 inches.

14 A weather balloon in the shape of a sphere has a diameter of 20 feet. What volume of helium gas can it hold? If each cubic foot of gas has a lifting power of 2 ounces, what is the maximum weight the balloon can lift?

15 Consider the two figures. What is the area of the base of each? What is the height of each? What is the volume of each? If each solid were cut by a plane parallel to the base at one-half its height, what would be the areas of the cross sections?

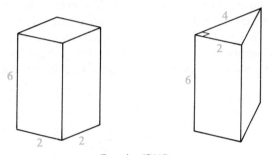

Exercise 6B(15)

6.3 PYRAMIDS AND CONES

The pyramid and cone are solids in which the measure of every cross section is different from the measure of the base. One approach to finding the volumes of such figures requires a principle first stated by an Italian mathematician named Cavalieri.

Cavalieri's principle. If two solids are included between the same two parallel planes, and if any plane parallel to these planes cuts cross sections of equal area in the solids, then the volumes of the two solids are equal.

This principle applies to pyramids of different shapes or even to different kinds of figures, such as the pyramids and cone shown in Illus. 6.12. Plane P_1 is parallel to planes P_2 and P_3. If the areas of the three cross sections C_1, C_2, and C_3 in plane P_1 are equal and the three figures have the same height, then the volumes of the three figures are the same. You should be able to see from Illus. 6.12 that if two solids are included between parallel planes, if the areas of the bases of the solids are equal, and

187

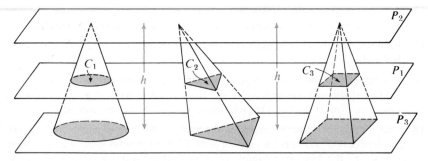

ILLUSTRATION 6.12   Three solids with equal cross section and equal volume

if the areas of the cross sections of the solids in every plane parallel to the bases are also equal, then the solids must have equal volumes.

Now what is needed to develop a formula is a relationship between prisms and pyramids. If we are given $ABFCDE$, a right triangular prism, with $\overline{ED} \perp \overline{CD}$, and we construct diagonals $\overline{CF}$, $\overline{DF}$, and $\overline{CB}$ (Illus. 6.13),

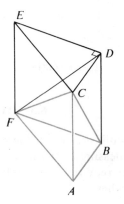

ILLUSTRATION 6.13   The volume of a pyramid is one-third the volume of the prism with the same base

three pyramids are formed. $ABFC$ is a pyramid with base ABF and with height equal to the measure of $\overline{CA}$; and $DECF$ is a pyramid with base CDE and with height equal to $\overline{EF}$. Since $\triangle ABF \cong \triangle CDE$ and $m(\overline{CA}) = m(\overline{EF})$, pyramid $A BFC$ and pyramid $DECF$ are congruent pyramids and have the same volume.

We can consider pyramid $CDEF$ as shown in Illus. 6.13 as having $\triangle EDF$ as base. Observe that this base is congruent to $\triangle DFB$, the other half of the rectangle $BDEF$. Pyramid $CDEF$ and pyramid $DBFC$ both

188

ILLUSTRATION 6.14   Frustums of cones and a pyramid

have the same height, $m(\overline{CD})$. Thus, these two pyramids have the same volume. Hence all three pyramids have the same volume. Since the volume of the prism is $V = A_b h$, and the pyramids can be represented by pyramid *ABFC* (which has the same base as the prism), we have $3V_{ABFC} = V_{prism}$, or $V_{ABFC} = \frac{1}{3}V_{prism}$. This is usually written as follows:

Formula 6.7   The volume V of a pyramid with a base of area A_b is $V = \frac{1}{3}A_b h$, where h is the height of the pyramid.

The justification given here was for a right pyramid with a right triangular base. This is a special case. However, a similar, though more complicated, proof can be made for the general pyramid, regardless of the shape of the base or the sizes of its angles. *Formula 6.7 applies to all pyramids.*

Cavalieri's principle allows us to write the formula for the volume of a cone immediately:

Formula 6.8   The volume V of a cone is $V = \frac{1}{3}A_b h$, where h is the height of the cone and A_b is the area of its base.

Two other solids, closely related to the pyramid and cone, are worth considering here. These are the *frustum of a pyramid* and the *frustum of a cone* shown in Illus. 6.14. Such a solid has similar bases that lie in parallel planes. Its volume can be found using the formula already developed. Consider the frustum of the right circular cone in Illus. 6.15. The volume

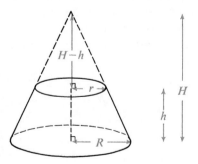

ILLUSTRATION 6.15   Finding the volume of the frustum of a cone

189

of the frustum is the volume of the whole cone, less the volume of the small cone which has been removed. Since the former has base area equal to πR^2, its volume is $\frac{1}{3}\pi R^2 H$. The height of the smaller cone is $H - h$, so its volume is $\frac{1}{3}\pi r^2 (H - h)$. The volume of the frustum is then $V = \frac{1}{3}\pi R^2 H - \frac{1}{3}\pi r^2 (H - h) = \frac{1}{3}\pi (R^2 H - r^2 H + r^2 h)$.

From similar triangles we observe that $(H - h)/r = H/R$. This yields $H = hR/(R - r)$. Substituting this in the equation for V we get

$$V = \frac{1}{3}\pi \left[R^2 \left(\frac{hR}{R - r} \right) - r^2 \left(\frac{hR}{R - r} \right) + r^2 h \right]$$

$$= \frac{1}{3}\pi \left[\frac{R^3 h}{R - r} - \frac{r^2 R h}{R - r} + \frac{r^2 R h - r^3 h}{R - r} \right]$$

$$= \frac{1}{3}\pi h \left[\frac{R^3 - r^3}{R - r} \right]$$

$$= \frac{1}{3}\pi h (R^2 + Rr + r^2)$$

Formula 6.9 The volume V of the frustum of a right circular cone is $V = \frac{1}{3}\pi h (R^2 + Rr + r^2)$, where R and r are the measures of the radii of the bases and h is its height.

The volume of the frustum of a pyramid poses a different problem, since rectangular or triangular bases are common. Without a regular polygon for a base, each case requires special consideration. However, if the pyramid has a square base the problem can be solved in a way analogous to that used above.

Formula 6.10 The volume V of the frustum of a right pyramid with square base is $V = \frac{1}{3}h(E^2 + Ee + e^2)$, where h is its height and E and e are the measures of the edges of its bases.

In the general case, the volume of a frustum can always be computed by sectioning it into a series of prisms and pyramids. The right rectangular frustum in Illus. 6.16 has been sectioned into nine parts: a right rectangular solid, four triangular prisms (two congruent pairs), and four

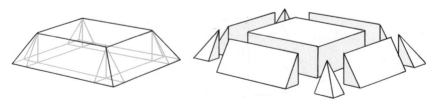

ILLUSTRATION 6.16   Partitioning a solid to find volume

190

congruent pyramids. The volume of the solid is the sum of the volumes of the nine solids.

Many polygonal solids that do not fit the formulas can be partitioned into several solids which do yield to formulas.

EXPERIMENT 6C

1 Suggest a means for determining by experimentation that the volume of a pyramid is one-third that of a prism with the same base area and height.

2 Repeat Experiment 6C(1) for a cone and cylinder under the same conditions.

3 Use the models to form a cone and a cylinder. How do their radii and heights relate? Using sand, beans, or some other convenient material, fill the cone and transfer the material to the cylinder. How many times can you do this before you fill the cylinder?

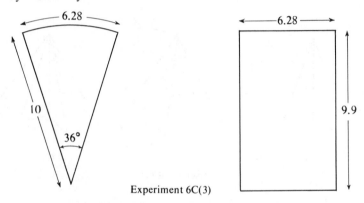

Experiment 6C(3)

4 Compare two cones with the same height and with radii in the ratio of 2:1. What is the ratio of their volumes?

5 Repeat Experiment 6C(3) with cones having the same base radius and heights in the ratio of 2:1.

6 Considering only right pyramids with regular bases, fill in the following table and then see if you can determine the relationship among the volumes.

No. of sides of base	Dimension of one side of base, inches	Height, inches	Volume
3	10	10	
4	10	10	
6	10	10	
8	10	10	

7 Obtain a measured volume (1 pint or 1 quart) of a substance such as sugar or sand. Note the number of cubic inches in the volume. Now pour the substance

191

out so that a neat pile is formed. Measure the diameter of the base of the pile and its height. Since you know the volume, compare your measurements to the formula for the volume of a cone. Repeat for two other quantities.

EXERCISE 6C

1 Calculate the volume of each of the figures.

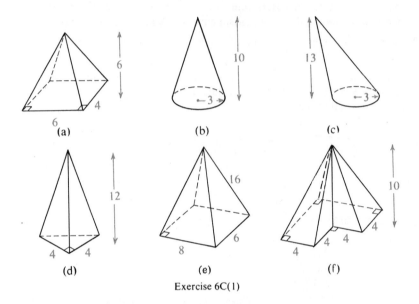

(a) (b) (c)

(d) (e) (f)

Exercise 6C(1)

2 Calculate the volume of a right circular cone with a radius of measure 10 inches and $h = 25$ inches. Calculate the volume of a square pyramid with $e = 10$ inches and $h = 25$ inches. Compare the volumes.

3 Find the volume of a bucket in the form of a frustum of a cone with the dimension shown. How many gallons does it hold?

Exercise 6C(3)

4 A pile of sand is in the shape of a circular cone. The circumference of the

192

base is 45 feet, and the pile is 5 feet high. How many cubic feet does it hold? How many cubic yards?

5   Would it be possible to load all the sand referred to in Exercise 6C(4) into a truck with a bed that is 6 feet by 10 feet by 4 feet?

6   Find the volume of each figure.

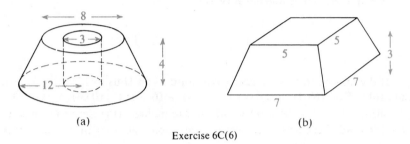

(a)   (b)

Exercise 6C(6)

7   How many cubic yards of earth would be required to fill 3000 flower pots shaped as frustums of right circular cones with inside dimensions as follows: upper diameter 10 inches, lower diameter 6 inches, height 10 inches?

8   A regular octahedron has edges with measure 3 inches each. Find its volume.

9   An ice cream cone has $h = 5$ inches and diameter of the base $2\frac{1}{2}$ inches. If the cone is completely filled with ice cream, and if a half-sphere of ice cream is put on top, how many cubic inches of ice cream have been used?

10   A second type of ice cream cone is shown in the illustration. How many cubic inches of ice cream will it hold if it is filled and then topped with a half sphere?

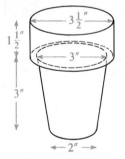

Exercise 6C(10)

6.4   SURFACE AREAS OF SOLIDS

Finding the surface area of a polyhedron requires no new development. Since we already know how to find the areas of polygons, and since polyhedra have polygonal faces, the total surface area can be found by cal-

193

culating the sum of the measures of all the faces. Several polyhedra that have been discussed before are of sufficient importance to warrant development of special formulas. The reader may note that in the development of surface-area formulas it becomes important to distinguish between *base area* and *lateral area*.

We first introduce a general formula.

Formula 6.11 Let S be the surface area of any polyhedron, and let $A_1, A_2, A_3, \ldots, A_n$ be the areas of the n faces of the polyhedron. Then the surface area $S = A_1 + A_2 + \cdots + A_n$.

If the polyhedron is a rectangular solid as in Illus. 6.17, then by Formula 6.11, $S = (6 \cdot 4) + (10 \cdot 4) + (10 \cdot 6) + (6 \cdot 4) + (10 \cdot 4) + (10 \cdot 6) = 24 + 40 + 60 + 24 + 40 + 60 = 248$ square inches. It is clear that in a rectangular solid there are three sets of congruent faces so that the surface-area formula could be written

$$S = 2(6 \cdot 4) + (10 \cdot 4) + 10 \cdot 6) = 2(24 + 40 + 60)$$

$$= 248 \text{ square inches}$$

Since this method is independent of the measure of the length of the sides, we may generalize.

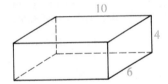

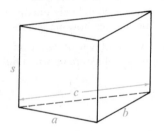

ILLUSTRATION 6.17   The surface area of the prism is 248 square inches

ILLUSTRATION 6.18   The surface area of a triangular prism is the sum of five areas.

Formula 6.12 Let l, w, and h be the length, width, and height of any right rectangular solid. Its surface area is

$$2lw + 2lh + 2wh = 2(lw + lh + wh)$$

In the special case of a cube, where l, r, and h all have the same measure, say e, the surface area is $S = 6e^2$.

Both the rectangular solid and the cube are special cases of the prism; a formula for the surface area of a prism will also apply to these two special figures. Consider the right triangular solid in Illus. 6.18. Five areas must be considered; two of them are areas of triangular bases, and the other three are the areas of the lateral faces that are rectangles. The rectangles all have a side with measure s; hence their areas are sa, sb, and sc. The total surface area of the solid can then be expressed as $S = 2A_b + s(a + b + c) = 2A_b + sP$, where P is the perimeter of the base. In the exercises we will investigate this formula, which applies to any right prismatic solid.

Formula 6.13 Given any right prismatic solid with height h and perimeter P, the surface area is $S = 2A_b + hP$, where A_b is the area of the base.

Formula 6.14 A right regular prism of n faces with height h and each side of measure e has surface area $S = 2A_b + hne$.

If the solid in Illus. 6.18 had not been a right prismatic solid, but rather appeared as in Illus. 6.19(a), we might question whether it would still be possible for us to find the measure of all the lateral surfaces. If the

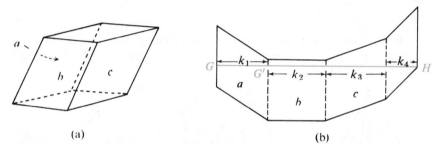

(a)   (b)

ILLUSTRATION 6.19   The surface area of the prismatic solid is the perimeter of a right section times the length of a slant edge

bases are removed we can imagine unfolding the lateral surfaces into a plane as in Illus. 6.19(b). We could then draw a line segment $\overline{GH}$ on this surface perpendicular to one edge. Since the edges of the figure are parallel, if $\overline{GH}$ is perpendicular to one edge it is perpendicular to all of them. Assume the vertical edges all have measure s. Since the measure of $\overline{GG'}$ is k_1, and this segment is the altitude of the first parallelogram, the area of this parallelogram is sk_1. Similarly, the areas of the other parallelograms are sk_2, sk_3, and sk_4. Remember $k_1 + k_2 + k_3 + k_4$ is the measure of the line segment $\overline{GH}$. If we reassembled the flattened figure in Illus. 6.19(b), the prismatic surface formed would have the segment $\overline{GH}$ on it. This segment defines a *right section* of the prism, that is, a section formed by a plane perpendicular to one of the lateral edges. This suggests a general formula:

195

Formula 6.15 Given any prismatic solid with a vertical edge, or slanted edge, of measure s and with right section with perimeter P, its surface area is $S = 2A_b + sP$, where A_b is the area of the base.

Finding the surface area of a pyramid requires a bit more ingenuity. If you keep in mind that all the lateral faces of a pyramid are triangles, then the problem is not difficult. In Illus. 6.20 an oblique pyramid and a right rectangular pyramid are shown. In both cases the height of the pyramid is h. However, this is not the height of each triangular face, so h cannot be used directly in a formula. What is needed is the measure k shown for one triangle in Illus. 6.20(a). This k can be different for each face, so a useful generalized formula cannot be developed. We can only state:

Formula 6.16 The surface area of a pyramid is

$$S = A_b + \tfrac{1}{2}(a_1 k_1 + a_2 k_2 + \cdots + a_n k_n)$$

where A_b is the area of the base; $a_1, a_2, a_3, \ldots, a_n$ are the measures of the sides of the base, and $k_1, k_2, \ldots, k_n$ are the measures of the corresponding altitudes of the lateral faces.

In the case of a regular pyramid like that in Illus. 6.21, the slant height is the same in every case. The area of a lateral face is $\tfrac{1}{2}sb$, where b

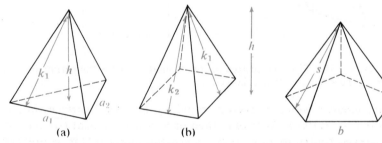

(a) (b) b

ILLUSTRATION 6.20 Finding the surface area of a pyramid

ILLUSTRATION 6.21 A regular pyramid of five faces

is the measure of one edge of the polygon. Since there are n congruent triangles, each with an edge of measure b, the lateral area is $\tfrac{1}{2}snb$. Since the base edges all have equal measure, nb is the perimeter of the base polygon.

Formula 6.17 The surface area of a regular pyramid is $S = A_b + \tfrac{1}{2}sP$ where A_b is the base area and s and P are the slant height and perimeter, respectively.

196

We now turn to the problem of finding surface areas of curved surfaces. We shall discuss three here: the cylinder, the cone, and the sphere.

Finding the surface area of a cylinder is not difficult when it is considered as the limiting case of a prism. Formula 6.13 applies to a right circular cylinder; in that case $P = 2\pi r$ and $A_b = \pi r^2$.

Formula 6.18   The surface area of a right circular cylinder is $S = 2\pi r^2 + 2\pi rh$, where r is the measure of the radius of the base and h is the height of the cylinder. This is sometimes written as $S = 2\pi r(r + h)$

Illustration 6.22 shows the pattern for the surface of a right circular cylinder. Does this suggest a reason why many containers are made in cylindrical shape?

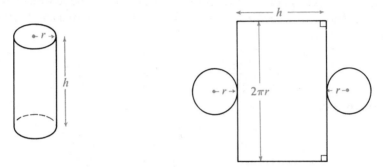

ILLUSTRATION 6.22   A pattern for a right circular cylinder

The surface of a cone can also be "unrolled" to lie flat in a plane (Illus. 6.23). Obviously finding the area of the lateral surface poses the difficult part of the problem. Illustration 6.24 shows the lateral surface as a sector of a circle with a radius of measure s, where s is the slant height of the cone ($s = \sqrt{h^2 + r^2}$). This area is readily computed by using proportionals.

ILLUSTRATION 6.23   A pattern for a right circular cone

197

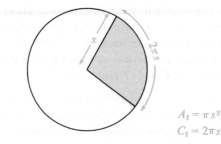

$$A_t = \pi s^2$$
$$C_t = 2\pi s$$

ILLUSTRATION 6.24   Finding the lateral area of a cone

In a circle with radius of measure s let the total area be represented by A_t. The area A_s of a sector is proportional to the total area in the same ratio as their arcs. Thus

$$\frac{A_t}{A_s} = \frac{2\pi s}{2\pi r} \qquad \text{or} \qquad \frac{\pi s^2}{A_s} = \frac{2\pi s}{2\pi r}$$

This resolves into $A_s = \pi rs$.

Formula 6.19   The surface area of a right circular cone is $S = \pi r^2 + \pi rs = \pi r(r + s)$, where r is the measure of the radius of the base and s is the slant height. If the height h is known, then $S = \pi r(r + \sqrt{h^2 + r^2})$.

There is no way to represent the surface of a sphere on a flat surface or with portions of a plane. This means, for example, that all maps (except globes) are distortions of the surface of the earth. The smaller the portion of surface to be represented, the better the plane representation will be, but some error will be introduced.

This means we must approximate the surface area of a sphere by geometric means or use mathematics beyond the scope of this course. Again we will simply state that the limit process of the calculus leads to an exact formula:

Formula 6.20   The surface area of a sphere is $S = 4\pi r^2$, where r is the measure of the radius of the sphere.

Methods of approximation will be given in the experiments.

EXPERIMENT 6D

1   Consider a 12-inch by 12-inch piece of sheet metal that is to be used to make a pan by cutting out the corners and turning up the edges. Fill in the following table to determine volumes:

198

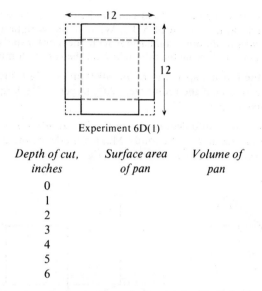

Experiment 6D(1)

Depth of cut, inches	Surface area of pan	Volume of pan
0		
1		
2		
3		
4		
5		
6		

At what cut dimensions is maximum volume obtained? Calculate the ratio of surface area to volume. When is this highest? Lowest?

2 Suppose you wish to enclose 1 cubic foot of volume in each of the following figures. Determine the required dimensions, and then calculate the surface area:

Figure	Dimensions	Surface area
Rectangular solid with $l = 2w$ and $w = h$:		
Cube:		
Cylinder with $r = h$:		
Cylinder with $d = h$:		
Sphere:		

3 Do the results of Experiment 6D(2) give you any indications about the minimum surface area required to enclose a given volume? About the solid that contains maximum volume in minimum surface area? If so, state a conjecture.

4 Suppose you are given a fixed amount of material, say 100 square inches, and you can mold it into any desired shape to enclose a volume. Which of the following solids gives maximum volume for a surface area of 100 square inches?

(a) Rectangular solid with $l = 2w, 2w = h$ (b) Cube

(c) Cylinder with $r = h$ (d) Cylinder with $d = h$

(e) Cone with $d = h$ (f) Frustum of a cone with $r = h$ and $R = 2r$.

(g) Sphere

Calculate the volume in each case.

199

5   Measure several cylindrical food cans and calculate the volume and surface area of each. Make a ratio of $S:V$. What is the variation obtained? Were there any cans of different dimensions with the same ratio of $S:V$? What is the minimum ratio? Note the ratio of $d:h$ on the can with minimum $S:V$ ratio.

6   Consider the Earth a sphere of radius 4000 miles. How much greater would the surface area be if the radius were 20% larger? 50% larger? 100% larger? Can you generalize?

7   When figure (a) is unfolded, figure (b) results. If $m(\overline{GH}) = 11$ inches, what is the lateral surface area of the solid? Make a model of the unfolded figure and refold. Do you see that $\overline{GH}$ determines a right section?

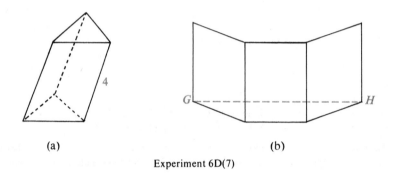

(a)   (b)

Experiment 6D(7)

8   How could you demonstrate that Formula 6.18 would be correct even if the cylinder was not a right cylinder? Give a deductive proof using Cavalieri's principle.

EXERCISE 6D

1   Develop the formula for the surface area of a cube from Formula 6.12.

2   Show that Formula 6.12 is a special case of Formula 6.13.

3   Show how Formula 6.14 follows from Formula 6.13.

4   Does it make any difference in the lateral area of a prism if the prism is a right prism, all other conditions being the same?

5   Develop a formula for the surface area of a prism, omitting one of the bases.

6   Derive the formula for the surface area of a right rectangular pyramid using the Pythagorean theorem and Formula 6.16.

7   Derive a formula for the surface area of a right pyramid with an equilateral triangle for a base.

8   Devise a method for finding the surface area of the frustum of a pyramid shown on the next page.

200

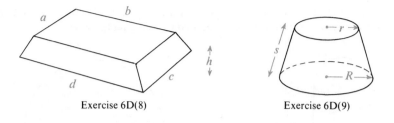

Exercise 6D(8)   Exercise 6D(9)

9 Devise a method for finding the surface area of the frustum of a cone shown.

10 A pile of wheat is in the shape of a cone. It has a diameter at the base of 20 feet and is 10 feet high. Find its lateral area. After a rainstorm it was found that water had penetrated to a depth of 2 inches in the pile. How many cubic feet of wheat were ruined? What fractional part of the total pile is this?

11 Find the surface area and volume of each of these figures.

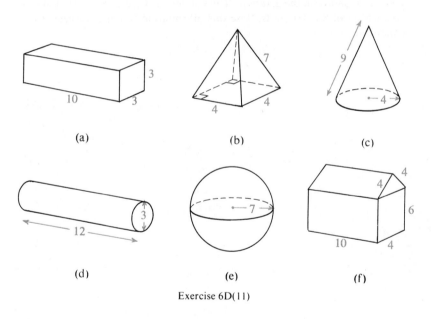

(a)   (b)   (c)

(d)   (e)   (f)

Exercise 6D(11)

12 A can manufacturer wishes to make a cylindrical can with a base radius of measure 2 inches. It must hold a quart. What must its height be?

13 A cylindrical tank is 12 feet high and 20 feet in diameter. The tank is to be painted except for the bottom. If 1 gallon of paint covers 200 square feet and costs $10, calculate the cost of the paint needed to cover the tank.

201

14 A pressure tank is shaped as shown, with hemispherical ends. Calculate the number of square feet of material needed to construct the tank.

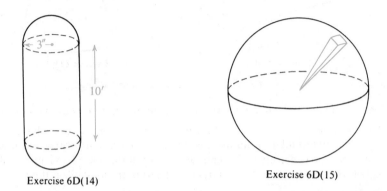

Exercise 6D(14)   Exercise 6D(15)

15 A sphere can be thought of as an infinite number of congruent pyramids, one of which is shown in the diagram. The volume of the sphere is the sum of the volumes of all the pyramids. Use this information to find a formula for the volume of a sphere.

202

7 / SYSTEMS OF MEASURE AND APPROXIMATION

7.1 STANDARD UNITS OF MEASURE

In preceding chapters we referred to basic units of measure and used such units as inches and feet, with no explanation of how these units were developed or how they relate to each other. And in Chapter 5 we created our own unit, the handy. The history of measurement shows clearly that many different basic units have been in use and that the choice of a basic unit is arbitrary. It can be considered an axiom of society that the more advanced a civilization is, the more precise and well defined its measuring system must be. Thus, present-day society demands much of a measuring system.

Two distinct measuring systems and two basic sets of measuring units are in use in the United States today; these are the *English* system and the *metric* system. They are also the two most widely used systems in the world. Although the English system is used quite generally in this country, the metric system is more popular elsewhere. It is interesting to note, however, that the metric system is the only system of measures ratified as a legal system of measures by the United States Congress (1866).

The English system is prevalent only in England, countries of the Commonwealth, and former English colonies. This system grew in the United States as a result of the English heritage of many of our founding colonies. However, owing to the obvious advantages of the metric system, and to increasing trade with countries who use the metric system, we are becoming more involved with this system. Almost all medicines are dispensed by metric measure; practically all scientific work is done in the metric system; Many grocery items have dual markings of weight; the military services measure distances or dimensions in meters and millimeters; international track meets are conducted with distances measured in meters.

How did it all start? Why do we have two systems? Let us begin by examining the English system. It appears that this system just "grew up" in England as commerce increased. Through the Middle Ages, a need for a standard unit of measure was recognized. It was indeed confusing to find that a pound in one village was not a pound in the next! Various local communities of England devised units which they used as standards. One of these, called the *inch*, was described as the length of three barleycorns laid end to end. This definition would lead one to believe inches were shorter in poor crop years than in good ones! A second standard measure decreed by a king for national acceptance was the *rod*. It was described as the total length covered by the feet of 20 townspeople standing toe to heel.*

At the time England was struggling with its new "standard" measures, the rest of the western world had even more poorly defined units of measure; but the growing pains of commerce were not as acute in these countries since trade was not as well developed in them. By the time the need for standardization was seen in European countries, England had adopted her own standards. In 1790 the metric system had its start; it was officially adopted by France in 1795. Since then, countries adopting standard measures have chosen the metric system.

Standardization of the English system in the Middle Ages made for uniform units of measure, but did not simplify the system. Thus, the English system is filled with unwieldy units with relationships such as:

$$
\begin{array}{rll}
43,560 & \text{square feet} & = 1 \text{ acre} \\
640 & \text{acres} & = 1 \text{ square mile} \\
8 & \text{pints} & = 1 \text{ gallon} \\
31\tfrac{1}{2} & \text{gallons} & = 1 \text{ barrel} \\
5\tfrac{1}{2} & \text{yards} & = 1 \text{ rod} \\
2,000 & \text{pounds} & = 1 \text{ short ton} \\
2,240 & \text{pounds} & = 1 \text{ long ton}
\end{array}
$$

The metric system is legal (but not mandatory) in the United States. Official conversion figures are kept in the archives of the U.S. Bureau of

---

*A history of the evolution of the English system of measure is interesting reading. The curious reader might refer to John Perry, *The Story of Standards*, Funk and Wagnall, New York, 1955.

Standards, and copies of conversion tables are readily available in reference books. The official meter for many years was designated as the distance between two scratches on a metal bar kept under guard in France in controlled temperature and humidity. Copies of this bar were made available, and one is kept by the United States government in Washington, D. C. Since there is apt to be some error in using any metal bar as a standard, and since it is possible that the standard could be lost or damaged, in recent years the meter has been defined in terms of the wavelength of a certain color of light. However, although the defining medium has changed, the measure of length has not. The yard is defined as 3600/3937 parts of a meter. Our other linear measures are derived from the yard. A comparison of these measures and conversions is given in Table 7.1.*

The table is by no means complete, particularly in the English system. There are many other measures, used in specific areas of commerce such as the dram, gill, hand, chain, etc.

An examination of Table 7.1 shows that it is simple to convert from one metric unit to the next. All units are multiples of 10 (or powers of 10). If units of measures are expressed in decimal form, then a change of decimal point is all that is needed to convert from one unit of measure to the next. This is not the case with the English system.

The metric system has none of the haphazard assignment of values of the English system, because the metric system was scientifically devised. The linear measure was decided upon first. The Earth itself was the yardstick (or meterstick), and multiples of 10 were the key. The distance from the North Pole to the Equator was measured and repeatedly subdivided by 10 until a unit of convenient size was found. The measure of length which seemed most suitable, called the *meter*, was one-ten-millionth of the designated distance.† Since the kilometer is 1000 meters, the distance from Pole to Equator is 10,000 kilometers. The circumference of the Earth has a measure of approximately 40,000 kilometers. The square and cubic measures in the metric system are logical extensions of the linear measure.

Some examples will demonstrate the computational advantages of the metric system. As a simple case suppose we are to change 156 centimeters to meters. Since 1 meter = 100 centimeters, we divide by 100; hence 156 centimeters = 1.56 meters. In the English system if we want to convert

---

*Most standard dictionaries list these conversions under *metric system*.

†Later measurements proved the original measurement was in error and the meter is not actually $\frac{1}{10,000,000}$ of the designated distance. This does not change the meter; it just means that the distance from the North Pole to the Equator is not 10,000,000 meters. The difference is very small.

205

TABLE 7.1   Standard Units of Metric and English Measures

Metric Measure	English Measure

LINEAR

Standard: meter	*Standard: yard*
1 millimeter = $\frac{1}{1000}$ meter	12 inches = 1 foot
1 centimeter = $\frac{1}{100}$ meter	3 feet = 1 yard
1 decimeter = $\frac{1}{10}$ meter	$5\frac{1}{2}$ yards = 1 rod
1 dekameter = 10 meters	1760 yards = 1 mile
1 hectometer = 100 meters	5280 feet = 1 mile
1 kilometer =1000 meters	

AREA (SQUARE)

Standard: square meter	*Standard: square yard*
100 square millimeters = 1 square centimeter	9 square feet = 1 square yard
1,000,000 square meters = 1 square kilometer	144 square inches = 1 square foot
	43,560 square feet = 1 acre
	640 acres = 1 square mile

VOLUME (CUBIC)

Standard: liter	*Standard: cubic yard*
1000 milliliters = 1 liter	27 cubic feet = 1 cubic yard
1 milliliter = 1 cubic centimeter*	1728 cubic inches = 1 cubic foot
	230.9 cubic inches = 1 gallon

WEIGHT

Standard: gram	*Standard: pound*
1000 grams = 1 kilogram	16 ounces = 1 pound
1000 kilograms = 1 metric ton	2000 pounds = 1 short ton
	2240 pounds = 1 long ton

*This was the intention of the measurements, but it has been shown that there is a small error so this is not actually true. The difference is very small.

156 inches to yards, we could reason that since 12 inches = 1 foot, then $\frac{156}{12}$ = 13 feet, and then since 3 feet = 1 yard, $\frac{13}{3}$ = $4\frac{1}{3}$ yards. This procedure involved division by 12 and then by 3. Of course two divisions are not necessary. We could say 156 inches × $\frac{1}{36}$ = $4\frac{1}{3}$ yards. But how much easier it is to divide by 100 in the metric system!

The computational differences become more apparent in volume measures. In the metric system a container with a measure of 2936 cubic centimeters contains 2.936 liters. Correspondingly, a jar which has a volume of 2936 cubic inches contains $\frac{2936}{230.9}$ gallons.

Another important measure is that of weight. We accept weight as a fixed property of an object, but actually the weight of the object depends on its distance from the center of gravity of the attracting body. Recent space flights have emphasized this. What we actually are measuring is

206

gravitational attraction. Although the weight of an object varies widely under different gravitational attractions, the *mass* of a body, that is, the amount of material in it, remains essentially unchanged. It would seem then that perhaps we should have units of measure of mass, not of weight. In the English system the basic unit of mass measure is called the *pound*, and the basic unit of weight measure is also called the *pound*. The pound unit of weight measure is defined as the force of gravity at sea level on an object with a mass of 1 pound. There is no practical significance to differences between mass and weight anywhere on the Earth's surface.

The metric system measure of weight is based on the *gram*. This is related to linear measure by defining 1 gram to be the weight of 1 cubic centimeter of distilled (pure) water at a temperature of 4°C at sea level. Thus, 1 liter of water (1000 cubic centimeters) weighs 1 kilogram.

Table 7.2 shows the relationships between the two systems. Other conversions can be derived from these. As an example of how we use

TABLE 7.2   Relationship Between Metric and English Measures

1 inch	$\doteq$ 2.54	centimeters
1 meter	$\doteq$ 39.37	U.S. inches
1 kilometer	$\doteq$ 0.621	miles
1 liter	$\doteq$ 0.264	U.S. gallons
1 kilogram	$\doteq$ 2.205	U.S. pounds

this table suppose we want to change 10 liters to gallons. Since 1 liter $\doteq$ 0.264 gallons, 10 liters is approximately $10 \times 0.264 = 2.64$ gallons. Similarly, to change 4.4 pounds to kilograms, we note that 2.2 pounds $\doteq$ 1 kilogram, and therefore 4.4 pounds $\doteq$ 2 kilograms. These two examples are quite obvious, but now suppose we need to change 6.0 pounds to kilograms. This is not an even multiple of 2.2, but we do know the number of kilograms and pounds are proportional; hence

$$\frac{1 \text{ kilograms}}{x \text{ kilograms}} = \frac{2.2 \text{ pounds}}{6 \text{ pounds}}$$

or

$$2.2x \doteq 6 \quad \text{and} \quad x \doteq 2.7 \text{ kilograms}$$

EXPERIMENT 7A

1 Why do you think many measuring tools used in the United States (e.g., rulers) are marked with both metric and English systems?

2 Estimate the lengths of the following segments in centimeters and in inches; then measure after all guessing is done.
(a) _____
(b) _____
(c) _____
(d) _____
In which system were you most accurate with your estimates? Can you explain?

3 Explain how congruence of standards affects a person buying a yard of cloth in Seattle and another in New York.

4 Suggest one or more areas of commerce where standards are not set and a difference in measurement can result.

5 Guess at the linear measure of one dimension of an accessible object in the classroom, such as a tabletop or a bookcover. Have everyone in the class do this. How many were above and how many were below the correct measure of length? Add all the guesses together and divide by the number of guesses. Is this average better than any single guess?

6 Repeat Experiment 7A(5) for a measure of area.

7 Estimate the distance between two buildings on campus and then measure. How good is your estimate? Repeat for a measure of area.

8 Estimate the length and width of this page and then measure to obtain a value for area. Do this in both the English and metric systems. Now measure. How good were your estimates?

9 Pick out several ordinary objects such as a wristwatch, textbook, newspaper, empty soda bottle, etc., and guess at their weights in pounds or ounces. Which were you better at guessing, the lighter or heavier objects?

10 Repeat Experiment 7A(9), only try guessing in metric weights.

11 Have each class member estimate the weight of an object (such as a dictionary or heavy book). Have each write down his estimate. Find the average guess and compare to the true weight.

12 Repeat Experiment 7A(11) with a pencil and a much heavier object such as a table or desk.

13 Repeat Experiment 7A(11) using metric weights. Check your estimates by converting them to English system measures.

14 Can you summarize by making a statement about your estimating ability of measures of length, weight, and area as related to your familiarity with the system and your experience?

EXERCISE 7A

1 Using Tables 7.1 and 7.2, (a) express 1 meter in feet and in yards; (b) express 1 liter in pints, quarts, and cubic inches.

2 Using Tables 7.1 and 7.2, (a) express 1 inch in meters, centimeters, and millimeters; (b) express 1 gallon in liters and milliliters.

3 Which conversions were easier, those in Exercise 7A(1) or in Exercise 7A(2)? Do you see any advantage in the metric system?

4 Perform the indicated conversions:
   (a) 1 mile to feet   (b) 1 mile to inches
   (c) 1 kilometer to millimeters   (d) 1 meter to millimeters
   (e) 1 square yard to square inches   (f) 1 square meter to square centimeters
   (g) 1 cubic foot to gallons   (h) 1 cubic meter to liters
   (i) 1 gram to kilograms   (j) 1 pound to long tons
   (k) 1 stone to ounces

5 Perform the indicated conversions:
   (a) 1 meter   = ____yards   1 yard   = ____meters
   (b) 1 inch   = ____centimeters   1 centimeter   = ____inches
   (c) 1 mile   = ____kilometers   1 kilometer   = ____miles
   (d) 1 square yard = ____square meters 1 square meter = ____square yards
   (e) 1 metric ton = ____short tons   1 short ton   = ____metric tons
   (f) 1 pound   = ____kilograms   1 kilogram   = ____pounds
   (g) 1 liter   = ____quarts   1 quart   = ____liters
   (h) 1 milliliter = ____cubic inches 1 cubic inch   = ____milliliters
   (i) 1 gallon   = ____liters   1 liter   = ____gallons

6 In the Olympics, distances are measured in meters. Convert the standard race distances of 100, 200, 400, and 1000 meters to yards.

7 The 1500-meter race is sometimes referred to as the "metric mile." Convert this distance to yards and compare to the mile. Which is greater?

8 The measure of the diameter of the bullet for a European gun is 7 millimeters. Compare this to a United States gun bore (diameter of barrel) with measure 0.38 inches.

9 Explain how the congruence property is used in measuring a tabletop with dimensions of 32 inches by 24 inches.

10 How do you think the congruence property is involved in the process of manufacturing a yardstick (the official yardstick is kept in a vault in Washington, D. C.)?

11 Compare cubic centimeters to the teaspoon measure if 48 teaspoons = $\frac{1}{2}$ pint. How many cubic centimeters of medicine are there in 1 teaspoon?

12 Investigate the following English units of measure and find their relationship to those in Table 7.1.
   (a) Rod, furlong, chain, league (linear)
   (b) Palm, span, cubit, pace (miscellaneous)
   (c) Fathom, cable, nautical mile (marine measure)
   (d) Nail, quarter (cloth measure)
   (e) Nip, gill, hogshead (liquid measure)

209

(f) Peck, bushel, chaldron (dry volume)

(g) Grain, scruple, dram (apothecaries' weight)

7.2   OTHER MEASURES

Thus far we have discussed five kinds of measure: linear, angular, area, volume, and weight. These constitute the majority of measures used, and certainly include those related to geometric figures.

However, there are many other measures involved in our everyday living and commerce that involve the idea of a basic unit and can usually be related to linear measure. These vary widely from the measurement of temperature (a measure of heat energy in the air) to the cost of a watch (a measure of commercial value) to IQ (a measure of intelligence compared to the population norm).

Standards are set in all these areas, so there is agreement on the basic units of measure. Almost all these other measures can be converted to a linear scale, and hence a linear measure.

For example, an intelligence scale can be devised as shown in Illus. 7.1 to show the relative scores of a group of students on a given test. Thus, a score of IQ 116 is greater than an IQ of 93.

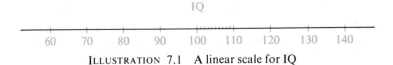

ILLUSTRATION 7.1   A linear scale for IQ

The reference point is 100 IQ points rather than 0.   The standard unit is 1 IQ point.   Though a device to measure such a quantity is an indirect measurement (the tool is applied to what the mind does rather than directly to the skull) we can express the results, such as they are, in linear units.

The measure of the temperature of the air involves a linear scale, which on Fahrenheit thermometers in common use ranges from − 50 units to + 120 units. On this scale 0 is arbitrary, as is the basic unit of 1 degree. On the centigrade scale for measuring temperature (used in conjunction with the metric system), 0° is the freezing point, and 100° is the boiling point, of water.

210

EXPERIMENT 7B

1   Investigate different standards of measure for each of the following. Decide if they are linear, describe the basic unit, and decide whether they are decimal in nature. Some may involve several systems of measure.
(a) Money   (b) Time
(c) Sound intensity   (d) Light intensity
(e) Temperature   (f) Electrical intensity
(g) Earthquake intensity   (h) Pressure
(i) Energy   (j) Speed
(k) Intelligence   (l) Educational level

2   Which of the above measures are less than exact; that is, apt to vary in time or place?

3   What does Experiment 7B(2) suggest in relation to principles of measurements?

EXERCISE 7B

1   Is the United States system of money a metric system? Compare it to the English system of pounds, shillings, and pence. Which is easier to convert in?

2   What principle of measurement is involved in the commercial value of a given object (such as a house) being $15,000 in Dallas and $18,000 in Chicago?

3   Using the conversion 1 centigrade degree = 1.8 Fahrenheit degrees, and $0°C = 32°F$, write a conversion formula. Then convert the following:
(a) $-50°C$ (b) $25°C$
(c) $120°C$ (d) $-40°F$
(e) $88°F$ (f) $134°F$

4   Many units of measure have a relationship to something that occurs naturally (i.e., $0°C$ is the freezing point of water, and $100°C$ is its boiling point). Find as many natural units of measure as possible in the first part of Experiment 7B.

5   Investigate the hardness scale for precious stones. How did this evolve? What is the standard unit? Is it linear?

6   Related to linear measures are clothing sizes. Determine the scale and the unit of measure for:
(a) Shoe size (men or women)   (b) Hat size (men)
(c) Glove size (women)   (d) Dress size (women)
(e) Shirt size (men)   (f) Coat size (men)
Is there a reasonable basis for most of these?

7.3   PRECISION

To find exactly how many panes of glass there are in a window you *count* the panes by saying "one, two, . . . ," and continue corresponding panes

211

and integers until you reach the last pane. Similarly, you can count *exactly* how many eggs there are in a carton, exactly how many fingers you have on one hand, and exactly how many students there are in a classroom. In each case it is easy to identify a single unit of the material being counted and consequently you can give an exact answer.

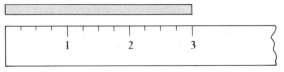

ILLUSTRATION 7.2   Measuring the length of a ribbon

If you are asked how long the ribbon in Illus. 7.2 is, it is not possible to give an exact answer. Let us see why. The ribbon is shown with a standard desk ruler below it. At first glance you might say that the end of the ribbon appears to be at the 3-inch mark on the ruler. If you move the book to the left and bring the book closer to your face, it will seem that the ribbon is more than 3 inches long. If you move the book to the right and close to your face the ribbon will appear to be less than 3 inches long. This illustration, although exaggerated, demonstrates one of the problems that always occurs in measuring by sight. The measurement is prejudiced by the way you look at things. You might say that you are looking directly down on the ribbon and the ruler and therefore the measure you give must be correct, but you have made no allowance for astigmatism, irregularly shaped eye surface, or atmospheric conditions.

A still more serious problem connected with the measure of an object such as the ribbon is the problem of precision. If we were to magnify the ribbon and the ruler we would be able to see that the end of the ribbon doesn't coincide with any of the marks of the ruler. We would need more marks between some of those existing to be able to read the length of the ribbon. Even with many more marks on the ruler there is no guarantee that the end of the ribbon would coincide with any mark. Since there is an infinite set of points between any two points there would have to be an infinite set of marks to guarantee that the end of the ribbon would always coincide with a mark. We cannot possibly construct such a ruler: consequently any measurement is inexact when applied to such an object. We can only give measurements correct to the smallest unit on the measuring device.

When you say you have a dozen apples you mean you have exactly 12 apples, no more or no less. However, when you say a stick has a measure

212

of 12 inches you mean only that it is near 12 inches in length. You may mean that it is between 11 and 13 inches in length, that it is between 11.5 and 12.5 inches in length, or that it is between 11.75 and 12.25 inches long. Illustration 7.3(a) shows a pencil stub and a ruler marked only in inch lengths. How long would you say the stub is? The same pencil is shown in Illus. 7.3(b) with a more closely defined ruler. Do you want to revise

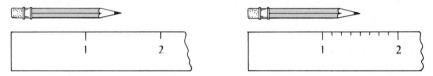

ILLUSTRATION 7.3   Finer markings aid in measuring

your estimate of the length of the pencil? You can see that the length you associate with an object depends primarily on how well defined a measuring device you use.

In Illus. 7.2 we suggested that the ribbon appeared to be about 3 inches long, but the line segment marking 3 inches on the ruler has width. Is a measure of exactly 3 inches at the beginning of the mark, in the middle, or somewhere else in that width? Most measuring devices will have line segments marking the divisions. No matter how fine these markings are, they have width and consequently will never allow an exact measurement. Generally, a measuring device with many subunits marked is considered to be superior to one with fewer markings. The smallest subunit marked on any measuring device is called the *least count* of that device.

Generally, measuring-device scales can be read to the nearest calibration mark. In Illus. 7.4 a segment is shown on a measuring scale which has a least count of 0.1. What measurement would you assign to the segment? Your answer should be 2.3 units. Does the endpoint appear to be closer to 2.3 on the scale than to 2.2 or 2.4? Your answer should be yes, for if it were closer to 2.2 we would assign the segment a measurement of 2.2 In this case, however, the endpoint lies between 2.25 and 2.35, and

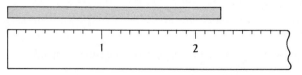

ILLUSTRATION 7.4   A ruler calibrated in tenths

213

consequently the measurement is given as 2.3. The *greatest possible error* is defined to be the greatest possible difference between the actual measurement and the measurement found by using the measuring device. In all common measuring devices the greatest possible error is one-half the least count.

On a ruler divided into 16 segments per inch, we agree that a measurement of $3\frac{1}{4}$ inches means a measure closer to $3\frac{4}{16}$ than to $3\frac{3}{16}$ or to $3\frac{5}{16}$. Similarly, a measure of $3\frac{1}{4}$ inches on a ruler divided into 8 segments per inch means a measure closer to $3\frac{1}{4}$ than to $3\frac{1}{8}$ or $3\frac{3}{8}$. The measurement of $3\frac{1}{4}$ inches on the ruler with 8 segments per inch is said to be *precise* to the nearest eighth.

A measurement written as 6.5 is interpreted to be precise to the nearest tenth, while a measurement of 3.42 is precise to the nearest hundredth. You should realize that there is a difference in recording a measurement as 6 or 6.0. The first notation indicates a measurement precise to the nearest unit, while the second indicates a measurement correct to the nearest tenth.

In our normal environment yet another error is introduced into measurement. The typical household has rulers and yardsticks which have been mass produced. The end of the stick may be uneven or shaved off, the stamping of the markings may not be properly placed, or the stick may be made of a material which expands and contracts readily. The standard foot is determined by the National Bureau of Standards. The marks used to produce the segmentation on ordinary rulers are copied from this standard.

With this awareness of the physical error in measuring devices and in the problems of reading a measurement so that it is even close to correct, you might question how measurement can be of use to us at all. We can use the best instruments available and the greatest of care, yet measurement is still inexact. It is fortunate that minor variations in readings of measurement can be tolerated in most cases and with care we can measure to a desired degree of accuracy.

EXPERIMENT 7C

1 Draw a line segment that you judge to be the same length as a dollar bill. Check your segment with an actual bill. Were you close?

2 Draw a line segment to represent the width of your text. Check your segment with the book.

3 Draw a line segment you estimate to have measure 3 inches. Check with a ruler. Were you close?

214

4 Close your eyes when the second hand on a watch is at 12. Open your eyes when you think 1 minute has elapsed. How close were you?

5 How many inches of barometric pressure does the mercury barometer in the drawing measure? What problem do you encounter in this reading? (The curve at the top of the mercury column is called a *miniscus*.)

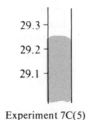

Experiment 7C(5)

6 Can you suggest a reason related to measurement that allows one mass-produced object (e.g., a washer) to work perfectly while another is "a lemon."

7 Measure a distance such as the width of a room several times (five or more). Find the average of these measurements. Have someone else do the same, and compare your results.

EXERCISE 7C

1 Name three physical limitations of the human body which would tend to cause error in reading a ruler.

2 Is it more useful to have a ruler divided into eights or into tenths? Explain your answer.

3 Is it more useful to have a ruler divided into tenths or hundredths?

4 The Taylor Park weather station in Colorado recorded a record low temperature of $-60°F$ on February 1, 1951. On January 18, 1943, the Island Park Dam station in Idaho recorded $-60°F$. Do you think the temperature was exactly the same in both places?

5 Why is it possible for the weight recorded on your bathroom scales to disagree with that recorded in the doctor's office?

6 The chains used in measuring distance in football are supposed to be 10 yards long. Give several reasons why this may not be true.

7 Which of the following give measurements correct to the nearest tenth? To the nearest hundredth?
(a) 4.8	(b) 6.19	(c) -3.17
(d) 0	(e) 208	(f) 0.032
(g) 5.06	(h) 6.00	(i) 0.16

215

8 For each of the ruler measurements given below, indicate the values between which the length should fall.

	Length recorded	Divisions on ruler
(a)	3.4	tenths
(b)	3.5	eighths
(c)	3.0	eighths
(d)	3.0	tenths
(e)	4	units
(f)	4.00	hundredths
(g)	4.0	sixteenths

9 A doctor says to a patient, "Your eyesight is 20/40." Are the numbers used exact or approximate?

10 A man says that his car can go 110 miles an hour. Is 110 exact? What would be necessary to determine his actual speed?

7.4 SIGNIFICANT DIGITS AND ROUNDING

In working with approximations it is important to understand the concept of *significant digits.* A significant digit is any digit used in writing a number except those zeroes which are used only for the purpose of locating the decimal point or those zeroes which do not have any non-zero digits on their left. This description assumes that there is at least one non-zero digit in the number. Thus, 641.07 has five significant digits, 0.004 has one significant digit, and 2001 has four significant digits. The number 27,000 may have as few as two or as many as five significant digits depending on whether the zeroes are used only to fix the decimal point or have real value. If 27,000 represents an approximation to the nearest thousand, then the zeroes are place holders and the number has two significant digits. However, if 27,000 represents the actual count of a set of objects, all five places are significant. The writer must indicate what is intended if we are to correctly assess the measure given as 27,000. The number 0.600 has three significant digits, as the zeroes are not used to locate the decimal but are used to indicate precision of measurement.

In working with numbers it is frequently convenient to *round* to a certain number of significant digits.

Definition 7.1 The number N is rounded to k significant digits when N is replaced by the number which is the closest approximation to N that can be written using k significant digits.

To round 57,846 to three significant digits we write 57,800 because 57,846 is closer to 57,800 than to any other number with three signifi-

216

cant digits. Similarly to round 1.576 to two significant digits we write 1.6 because 1.576 is closer to 1.6 than to any other number with two significant digits. In the event that a given number is exactly half-way between two numbers with the correct number of significant digits, we agree to round to the nearest *even* number. Therefore, 8150 rounded to two significant digits is 8200.

The number obtained by rounding depends upon the number of significant digits retained. Consequently, 78,462 rounded to four significant digits is 78,460; to three significant digits is 78,500; to two significant digits is 78,000; to one significant digit is 80,000.

Sometimes a situation requires that we round to the nearest hundredth, tenth, or unit rather than rounding to a specified number of significant digits.

Definition 7.2   A number *N* with significant digits to the right of the tenths (hundredths, thousandths, . . .) place is rounded to the nearest *tenth* (hundredth, thousandth, . . .) when *N* is replaced by the number which is the closest approximation to *N* which contains no digits to the right of the *tenths* (hundredths, thousandths, . . .) place.

The number 18,765.198 rounded to tenths is 18,765.2, as this is the closest approximation that can be made with no digits to the right of the tenths place. Similarly 987.21 rounded to units is 987, because this is the number closest to 987.21 which has no significant digits to the right of the units place.

EXPERIMENT 7D

1   Find examples of measures in technical magazines such as *Popular Mechanics* or *The Handyman's Guide.* How many significant digits are given in the measurements you find there? Are there any cases where you are in doubt?

2   Check an almanac, encyclopedia, or dictionary for the interplanetary distances in the solar system. How many significant digits are used?

3   Suggest three common measures (e.g., miles per hour) for which we use two significant digits; three; five. Can you find a measure for which more than five significant digits are used?

4   Change the measures in a recipe to decimal equivalents. How many significant digits did you use?

5   Measure the size of a room to the nearest quarter of an inch. How many significant digits did you use?

6   Repeat Experiment 7D(5) using the metric system. Which was easier to convert? Was the number of significant digits different?

217

7   Check several grocery items for the number of significant digits in their measures.

8   Check the financial page of a daily paper for monetary reports. How many significant digits are used?

EXERCISE 7D

1   Indicate the number of significant digits in each of the following:
  (a) 1.76   (b) 0.01764   (c) 17.6
  (d) 1005   (e) 8.00006   (f) 14
  (g) 0.0301   (h) 500   (i) 500.0
  (j) 0.000007   (k) 29,564.8   (l) 3.14159

2   Round each of the following numbers to four significant digits:
  (a) 9841.5   (b) 6000.52   (c) 0.00173452
  (d) 959,764   (e) 2700   (f) 96.979

3   In Exercise 7D(2), round each of the numbers to three significant digits; to two significant digits; to one significant digit.

4   Round each of the following numbers to tenths:
  (a) 472.136   (b) 418.759   (c) 10.09875
  (d) 961.765   (e) 0.00162   (f) 939.31426

5   In Exercise 7D(4), round each of the numbers to hundredths; to units; to tens; to hundreds.

6   Give an example of a number which has the same value rounded to hundreds and to tens.

7   Give an example of a number which has the same value when rounded to tens and units.

8   Does a number with three decimal places necessarily have three significant digits?

9   Does a number with three significant digits necessarily have three decimal places?

10 Give an example of a number which has three decimal places and three significant digits.

7.5 ABSOLUTE AND RELATIVE ERROR

Since we know that the numbers we use to report measurements are not exact we need some system for describing the error involved in any measurement. If you are told a bolt of cloth is 40 feet long and that the measurement is correct to the nearest 10 feet, then you know only that the cloth is between 35 and 45 feet long; if the measurement is correct to the nearest foot then you know the cloth is between 39.5 and 40.5 feet long.

Let us suppose that the measurement is correct to the nearest 10 feet; the report of the measurement may be in error by as much as 5 feet. In that case 5 feet is called the *absolute error* or *greatest possible error*.

Definition 7.3 The *absolute error* of a measurement is the greatest possible difference between the true measurement and the approximate measurement.

In this same example, the error is at most 5 feet in a measurement reported to be 40 feet; the error is 5 parts in 40 or $\frac{5}{40} = \frac{1}{8}$. One-eighth is called the *relative error*.

Definition 7.4 The *relative error* or *accuracy* of a measurement is the ratio of the absolute error to the approximate measurement.

If the measurement of 40 feet is correct to the nearest foot, then the absolute error is 0.5 foot (the actual length is between 39.5 and 40.5 feet), and the relative error is $\frac{0.5}{40} = \frac{5}{400} = \frac{1}{80}$. Of these two error measures, relative error is the more meaningful for comparisons. For example, suppose the length of a piece of metal tubing is reported as 4 feet, correct to the nearest foot. This means the pipe is between 3.5 and 4.5 feet long, and the absolute error is 0.5 feet. This is the same absolute error we had for the bolt of cloth. However, the relative error in the case of the tubing is $\frac{0.5}{4} = \frac{5}{40} = \frac{1}{8}$. This compares to the value $\frac{1}{80}$ computed for the bolt of cloth. That is, the error in measuring the pipe is 10 times the error in measuring the cloth.

EXPERIMENT 7E

Determine the absolute error and the relative error in each of the following situations.

1   A man buys 12.7 gallons of gas at a service station.

2   A carpenter measures a board to be 5 feet $4\frac{1}{2}$ inches.

3   A truck weighs 43,200 pounds.

4   A housewife buys 25 pounds of sugar.

5   An auto odometer registers 68.9 miles

6.   A savings account accrues $5.17 interest.

Perform the following measures and determine your relative and absolute errors.

7   Measure the dimensions of a textbook.

8   Measure the size of your room.

9   Measure the diameter of a nickel.

219

10   Weigh yourself.

11   Measure someone's height.

12   Measure the volume of a food can (use water and a calibrated cup).

13   Would any of your answers to Experiments 7E(7) to 7E(12) vary according to the measuring tool used?

EXERCISE 7E

1   Complete the following table:

Measure	(*Precision*) Absolute error	(*Accuracy*) Relative error
2.50 inches	0.005 inch	
400 feet	10 feet	
9500 miles		$\frac{1}{100}$
16 inches		$\frac{1}{16}$

2   A man reports that he expects his measurements to have a relative error of $\frac{1}{60}$. If he measures a sidewalk and reports it to be 120 feet long, how long might it actually be?

3   If in Exercise 7E(2) the man says he expects an absolute error of 7 feet, how long might the sidewalk be?

4   If a measurement of 40 feet has a relative error of $\frac{1}{20}$, what would you expect the absolute error to be in a measurement of 80 feet?

5   If a measurement of 90 feet has a relative error of $\frac{1}{10}$ what would you expect the absolute error to be in a measurement of 135 feet?

6.   If a measurement of 72 inches has an absolute error of 2 inches, what would you expect the relative error to be in a measurement of 36 inches?

7   In each pair of numbers, which has the largest relative error?
(a) 16.8 feet, 0.172 feet   (b) 3.8 pounds, 0.014 ounces
(c) 196 tons, 8.91 pounds   (d) 8 pints, 0.004 quarts

8   What are the absolute and relative error in each of the following?
(a) 691.5 feet   (b) 1.602 inches   (c) $2007
(d) 4.1685 grains   (e) 0.017 ounces   (f) 16,800 pounds

7.6   COMPUTING WITH APPROXIMATIONS TO NUMBERS

It should be obvious that if approximations to measurements are used in computations, then the results of the computations are also approximations. It is necessary to have some method for describing the errors and for adjusting answers in light of accumulated error.

Suppose the numbers 9.4 and 17.6, which represent approximations to measurements, are to be added. Since 9.4 lies between 9.35 and 9.45 and 17.6 lies between 17.55 and 17.65, we know the sum is between 26.90 (9.35 + 17.55) and 27.10 (9.45 + 17.65). Another way of stating this sum is as 27.0 ± 0.10. Similarly, the sum of 7.2 and 3.18 can be expressed as 10.38 ± 0.055. In this case the first number is between 7.15 and 7.25 while the second is between 3.175 and 3.185. Adding the smaller figure in each case gives 10.325 while the sum of the larger figures is 10.435; hence the result is 10.38 ± 0.055. Notice that in each case the absolute error of the result is the sum of the absolute errors of the numbers added.

Now, although this second example seems to be mathematically correct, there is an obvious error in the sum. Two measurements have been made, one of them precise to one decimal, the other precise to two. Although it may seem a little unusual to use two different measuring devices with different precisions, it is common to find two numbers like this. If one of the measurements is correct only to tenths then the sum of the measurements can be correct only to tenths. No matter how many measurements with precision of one-hundredth are added to the measurement with precision one-tenth, the sum is never more precise than the least precise measurement. We can summarize by stating.

The sum of two numbers with k and n significant digits contains k or n significant digits, whichever is smaller.

Let us suppose now that we measure the length and width of a room and record the measurements as 17.6 and 31.8 feet, respectively. The area of the room is their product. We can find three products associated with these measurements.

17.55	17.6	17.65
31.75	31.8	31.85
87 75	140 8	88 25
1228 5	176	1412 0
1755	528	1765
5265		6295
557.2125	559.68	562.1525

We can see from these calculations that the product lies between 557.2125 and 562.1525. If we consider 17.6 and 31.8 as exact numbers the product is 559.68. We are now faced with the problem of deciding what value to report for the product. For consistency we make the following rule:

The product of two numbers with k and n significant digits, respectively, is reported with the same number of significant digits as appear in the factor with the least number of significant digits.

221

You should be aware that products found using this rule may appear more exact than they really are. If we apply the rule to the example we see that both 17.6 and 31.8 have three significant digits, and consequently the product should have three significant digits; that is, we would write the product as 560. Our previous understanding of this notation would now lead us to interpret the number 560 as a number which lies between 559.5 and 560.5, yet our original calculations indicate that this is not necessarily true and that the number may be anywhere between 557.2125 and 562.1525.

The rule used for multiplication is also used for division. Consequently the quotient $\frac{4003}{25.2}$ is reported not as 158.849206, but rather as 159, a number correct to three significant digits.

EXPERIMENT 7F

1 Determine how error might accumulate in the following situations:
   (a) Jet fuel weighs 6.9 pounds per gallon; 5000 gallons of jet fuel are loaded into an aircraft.
   (b) The following parts are in a casting assembly for which you are to determine the weight: 8 bolts at 0.24 pound each; 22 rivets at 0.0003 pound each, 8 washers at 0.01 pound each, 8 nuts at 0.12 pound each, 1 gasket at 0.003 pound, and 2.4 pounds of cast metal.
   (c) A sheet of paper weighs 0.03 ounce; 326 pages are used in a report.
   (d) 212 suitcases are loaded aboard an aircraft, their weights determined to the nearest 5 pounds.
   (e) A salesman reports the following distances travelled on a series of trips: 120 miles, 50 miles, 200 miles, 320 miles, 85 miles, 16 miles, 4 miles, 160 miles, 100 miles, 5 miles.

2 Suggest some situations in which measurements of different precisions are encountered and the results of these measurements are to be combined.

3 Measure a room and determine its area. Determine the tolerances of your answer.

4 A strip of concrete pavement 32 feet wide, 8.1 miles long, and 12 inches thick is to be constructed. Determine the absolute error of each measurement if the length is to the nearest tenth of a mile, the thickness to the nearest inch, and the width to the nearest 3 inches.

EXERCISE 7F

1 What is the absolute error accumulated in each of the following?

(a)	17.6	(b)	0.00017	(c)	186.75
	421.9		1.07		−0.15
	+1687.92		+16.076		

222

2 Perform the following multiplications and divisions reporting the results according to the rule given in this section.
   (a) $(17.1)(6.235)$ (b) $\frac{98.5}{17}$
   (c) $(56.175)(0.30)(54.2)$ (d) $(0.004)(891.23)$
   (e) $\frac{0.0012}{18}$

3 An acre of land contains 43,560 square feet. A field is reported to be 745 feet long and 385 feet wide. What is the smallest possible number of acres (to the nearest hundredth) that can be in the field? The largest number?

4 A man has a measuring rod with marks at each yard. He reports the dimensions of a rug correct to the nearest yard to be 4 yards by 9 yards. What is the smallest possible area of the carpet? The largest?

5 In Exercise 7F(4), if the cost of the carpet is $10 per square yard, what would be the spread of costs?

6 A yard was at one time described as the length from the tip of one's nose to the end of one's index finger. Roughly, how many inches of error would result in various female adults measuring a yard.

7 State a rule concerning accuracy of the subtraction of two numbers found by measuring.

8 A computer can keep eight significant digits in storage for any single operation. It automatically selects the most significant digits. What would be the result in storage for each of these sums?
   (a) $187 + 2954.3$
   (b) $16,397 + 0.05$
   (c) $2,000,715 + 0.0008314$

9 Using the premises in Exercise 7F(8), what are the results of these subtractions?
   (a) $24,631 - 13.8$
   (b) $186,973 - 0.054$
   (c) $276,827.31 - 0.0000076$

10 Using the premises in Exercise 7F(8), what is the result stored for each multiplication?
   (a) 12×100
   (b) 200×2.54
   (c) 3000×0.000007

8 / ANALYTIC GEOMETRY—A DIFFERENT APPROACH

8.1 DIRECTED LINE SEGMENTS

Analytic geometry is concerned with the application of algebraic methods to geometric problems. In this chapter we will use the methods of analytic geometry to provide solutions for some problems posed in earlier chapters. Some of the solutions given earlier were based on inductive methods. However, since analytic geometry is based on algebraic techniques, the solutions given here will be deductive.

In Chapter 5 we discussed the number line. We selected an arbitrary point on a line like that in Illus. 8.1 and labelled it 0. We then picked a

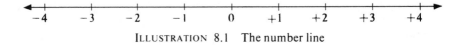

ILLUSTRATION 8.1   The number line

unit length and marked units to the left and right of zero, labelling them $\ldots, -3, -2, -1, 0, +1, +2, +3, \ldots$. Although not pictured here, there were points on the line that correspond to such numbers as $\frac{1}{2}$, $\frac{1}{4}$, $\sqrt{2}$, and $\sqrt{3}$. One problem in analytic geometry is finding a formula for the measure of the distance between two points on the number line. We call this distance a *directed distance*, because we consider one direction as positive and the opposite direction as negative. Directed distances measured to the right along the number line are positive, whereas those mea-

ILLUSTRATION 8.2   The directed distance from A to B is 6 units

sured to the left are negative. For a vertical line, distances measured up are positive; those measured down are negative. In Illus. 8.2 the measure of the directed distance from A to B is 6 units, whereas the measure of the directed distance from B to A is -6 units. We use the notation $m_d(\overline{AB})$ to indicate the measure of the directed distance from A to B. For any two points A and B on the number line, $m_d(\overline{AB}) = -m_d(\overline{BA})$.

If we assign a value to each directed line segment with one endpoint on O as shown in Illus. 8.3, we have $m_d(\overline{OA}) = x_1$ and $m_d(\overline{OB}) = x_2$. Then the formula for $m_d(\overline{AB})$ is

$$m_d(\overline{AB}) = m_d(\overline{OB}) + m_d(\overline{AO})$$

Since
$$m_d(\overline{AO}) = -m_d(\overline{OA}) = -x_1$$

we have
$$m_d(\overline{AB}) = x_2 - x_1$$

Similarly,
$$m_d(\overline{BA}) = m_d(\overline{BO}) + m_d(\overline{OA})$$

In this case
$$m_d(\overline{BO}) = -m_d(\overline{OB}) = -x_2$$

so that
$$m_d(\overline{BA}) = -x_2 + x_1 = x_1 - x_2$$

Although Illus. 8.3 shows both A and B to the right of O, this was not used in developing the formulas. The experiments will show that the conclusion is the same no matter where A and B are located.

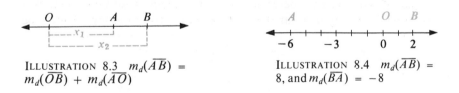

ILLUSTRATION 8.3 $m_d(\overline{AB}) =$ $m_d(\overline{OB}) + m_d(\overline{AO})$

ILLUSTRATION 8.4 $m_d(\overline{AB}) =$ 8, and $m_d(\overline{BA}) = -8$

As an example of how we might use these formulas, suppose that $m_d(\overline{OA}) = -6$ and $m_d(\overline{OB}) = 2$ as in Illus. 8.4. Then

$$m_d(\overline{AB}) = m_d(\overline{OB}) + m_d(\overline{AO}) = 2 + 6 = 8$$

and
$$m_d(\overline{BA}) = m_d(\overline{BO}) + m_d(\overline{OA}) = -2 - 6 = -8$$

225

The addition of measures of directed line segments is done algebraically. Consider three points A, B, and C on a line as in Illus. 8.5: $m_d(\overline{AB}) + m_d(\overline{BC}) = m_d(\overline{AC})$ for this case. But this is a special arrangement of A, B, and C. Another arrangement is shown in Illus. 8.6. In this case, $m_d(\overline{AB}) + m_d(\overline{BC}) = m_d(\overline{AC})$ again, since $m_d(\overline{BC}) = -m_d(\overline{CB})$ and $m_d(\overline{AB}) - m_d(\overline{CB}) = m_d(\overline{AC})$. Still another arrangement of A, B, and C is given in Illus. 8.7. In this case it is also true that $m_d(\overline{AB}) + m_d(\overline{BC}) =$

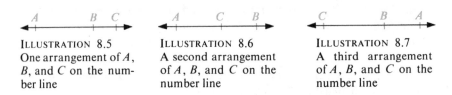

ILLUSTRATION 8.5
One arrangement of A, B, and C on the number line

ILLUSTRATION 8.6
A second arrangement of A, B, and C on the number line

ILLUSTRATION 8.7
A third arrangement of A, B, and C on the number line

$m_d(\overline{AC})$. From our last example we can see that $m_d(\overline{CA}) + m_d(\overline{AB}) = m_d(\overline{CB})$ but $m_d(\overline{CA}) = -m_d(\overline{AC})$ and $m_d(\overline{CB}) = -m_d(\overline{BC})$. Substituting these, we find $-m_d(\overline{AC}) + m_d(\overline{AB}) = -m_d(\overline{BC})$. Another way of writing this same equation is $m_d(\overline{AB}) + m_d(\overline{BC}) = m_d(\overline{AC})$. In each case we have investigated, the result was the same; namely, $m_d(\overline{AB}) + m_d(\overline{BC}) = m_d(\overline{AC})$. Although it would be premature for us to draw a conclusion before we have investigated all possible arrangements of A, B, and C on the line, our intuition suggests a formula. We will examine the remaining cases in the experiments but state the following theorem here.

THEOREM 8.1   *Given any three points A, B, and C on a line,*

$$m_d(\overline{AB}) + m_d(\overline{BC}) = m_d(\overline{AC}).$$

Many problems require more than measurement along a line. In Illus. 8.8 we have drawn a facsimile of a window with a small spider on one of the panes. How can we describe where the spider is located? One way would be to specify how far the spider is down from the top of the win-

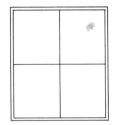

ILLUSTRATION 8.8   A spider on a window

dow and how far he is to the left of the right-hand side. He also can be located by stating how far he is to the right and how far he is above the mullion bars separating the panes of glass. No matter where the spider is on the window, we can locate it by stating its position in terms of some number of units to the right or left of the vertical bar and some units above or below the horizontal bar. This is a physical example of a coordinate system.

To establish a coordinate system, we construct two perpendicular number lines. Ordinarily, one of the lines is vertical and the other horizontal. The horizontal line, or axis, is usually designated the *x axis*, and the vertical line the *y axis*. The point at which the two lines intersect is called the *origin* and is assigned the reference number zero on both axes. As before, points to the right of zero on the horizontal axis are positive, and points to the left are negative. Convention also dictates that points on the vertical axis above the horizontal axis are positive, while those below are negative. Using these crossed lines, or *Cartesian coordinate system*, we can refer to any point in the plane. For example, point *P* in Illus. 8.9(a) is

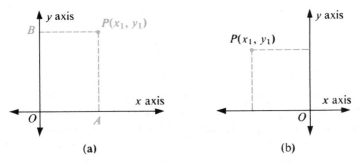

ILLUSTRATION 8.9   Locating points on the coordinate axes

determined by the intersection of two specific lines. One of these lines is parallel to the *y* axis, while the other is parallel to the *x* axis. As illustrated, the line parallel to the *y* axis cuts the *x* axis at *A*, and the line parallel to the *x* axis cuts the *y* axis at *B*. If $m_d(\overline{OA}) = x_1$ and $m_d(\overline{OB}) = y_1$, then the position of *P* can be given as (x_1, y_1). In Illus. 8.9(a) both x_1 and y_1 are positive. However, if x_1 is negative and y_1 is positive, then *P* would be positioned as shown in Illus. 8.9(b). The positioning numbers (x_1, y_1) are conventionally written in that order, and are called an *ordered pair*. The signed numbers x_1 and y_1 are the coordinates of *P*. The

227

x coordinate is called the *abscissa*, and the y coordinate the *ordinate*. In Illus. 8.10, P has an abscissa of 4 and an ordinate of -3; Q has co-ordinates $(-5, -1)$. The four portions of the coordinate system are called quadrants and are numbered as shown in Illus. 8.11.

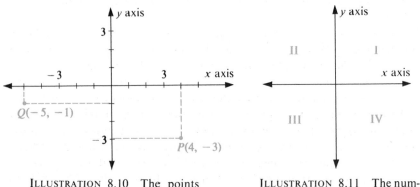

ILLUSTRATION 8.10 The points $Q(-5, -1)$ and $P(4, -3)$ on the coordinate axes

ILLUSTRATION 8.11 The numbering of the quadrants

EXPERIMENT 8A

1 Locate three points on a line labelling them C, B, and A, in order from left to right. Verify that $m_d(\overline{AB}) + m_d(\overline{BC}) = m_d(\overline{AC})$.

2 Locate three points on a line, labelling them B, A, and C, in order from left to right. Verify that $m_d(\overline{AB}) + m_d(\overline{BC}) = m_d(\overline{AC})$.

3 Locate three points on a line, labelling them B, C, and A in order from left to right. Verify that $m_d(\overline{AB}) + m_d(\overline{BC}) = m_d(\overline{AC})$.

4 Are there any arrangements of A, B, and C other than those given in Illus. 8.5, 8.6, and 8.7 and those you drew in Experiments 8A(1), 8A(2), and 8A(3)? Has Theorem 8.1 been established?

5 Locate A, B, and C in order on a vertical line. Does $m_d(\overline{AB}) + m_d(\overline{BC}) = m_d(\overline{AC})$?

6 In Experiment 8A(5) if A, B, and C are rearranged, is the result the same?

7 On a Cartesian coordinate system locate the points $R(3, 5)$, $S(-5, 0)$, $T(3, -1)$, $U(-2, -4)$, $V(-6, 2)$, $W(0, 0)$, $Z(3, -5)$.

8 Construct a rectangular coordinate system with positive values to the left of 0 and negative values to the right of 0 on the x axis, but with the y axis as it appears in the usual coordinate system. Locate the same points called for in Experiment 8A(7). Are any of the points the same? Do any points exist which would be the same in both experiments?

228

9 Construct a rectangular coordinate system with positive values to the left of 0 and negative values to the right of 0 on the *x* axis, and negative values above the *x* axis and positive values below the *x* axis on the *y* axis. Locate the points you located in Experiment 8A(7). Are any of the points the same as in Experiment 8A(7)? Do any points exist for which the plotting would be the same?

10 Consider the coordinate system shown, in which the axes are not perpendicular.
(a) Is it possible to locate every point in the plane using this coordinate system?
(b) In the Cartesian coordinate system described in the text, what is the value of α?
(c) Draw a system which has $\alpha = 60°$.
(d) Draw a system which has $\alpha = 10°$.
(e) In these systems locate the points given in Experiment 8A(7). Did you locate different points than those in Experiment 8A(7)?
(f) Compare the ease of using systems (c) and (d) to that of the Cartesian coordinate system.

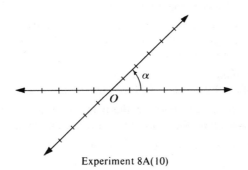

Experiment 8A(10)

EXERCISE 8A

1 On the number line, if $A = 4$, $B = 6$, and $C = 9$, find (a) $m_d(\overline{AC})$; (b) $m_d(\overline{BC})$; (c) $m_d(\overline{CA})$.

2 On the number line, if $A = -7$, $B = 2$, and $C = 5$, find (a) $m_d(\overline{AC})$; (b) $m_d(\overline{BC})$; (c) $m_d(\overline{CA})$.

3 On the number line, if $A = 6$, $B = 9$, and $C = -8$, find (a) $m_d(\overline{AC})$; (b) $m_d(\overline{BC})$; (c) $m_d(\overline{CA})$.

4 On the number line, if $A = -5$, $B = -7$, and $C = -2$, find (a) $m_d(\overline{AC})$; (b) $m_d(\overline{BC})$; (c) $m_d(\overline{CA})$.

5 Does the order of A, B, and C on a line ever make any difference in the use of the formula $m_d(\overline{AB}) + m_d(\overline{BC}) = m_d(\overline{AC})$?

229

6 Using the diagram as a guide and letting $m_d(\overline{OA}) = x_1$ and $m_d(\overline{OB}) = x_2$, show $m_d(\overline{AB}) = x_2 - x_1$ and $m_d(\overline{BA}) = x_1 - x_2$.

Exercise 8A(6)   Exercise 8A(7)

7 Using the diagram as a guide and letting $m_d(\overline{OA}) = x_1$, $m_d(\overline{OB}) = x_2$, and $m_d(\overline{OC}) = x_3$, prove algebraically that $m_d(\overline{AB}) + m_d(\overline{BC}) = m_d(\overline{AC})$.

8 Does your proof in Exercise 8A(7) depend upon the order of A, B, and C?

9 A rectangle is constructed with sides parallel to the axes and center at the origin. If one vertex is the point $A(4, 3)$, what are the coordinates of the other vertices?

10 A rectangle $ABCD$ has one side on the x axis and one vertex at $A(0, 0)$. Another vertex is at $B(5,4)$. Find the other two vertices.

11 What is the ordinate of any point on the x axis?

12 What is the abscissa of any point on the y axis?

13 One side of a square with an area of 16 square units is the line segment determined by the points $A(-3, 1)$ and $B(1, 1)$. Find the coordinates of the other vertices. Is there more than one answer?

14 Draw a right triangle with two vertices at $A(3, 5)$ and $B(-2, -2)$. Find the coordinates of the third vertex. Is there more than one possibility?

15 If the coordinates of any point in the plane were given as (y_1, x_1) rather than (x_1, y_1), would we still be able to refer to each point in the plane?

16 Demonstrate the position of $A(6, 4)$ and $B(2, -2)$ using the system in Exercise 8A(15).

8.2 DISTANCE BETWEEN TWO POINTS

In the last section we determined a formula for the directed distance between two points on a line. In many applications we are interested in the measure of the distance between two points but are not interested in the direction in which the measurement is made. The measure of the distance from A to B without regard to direction is denoted by $m(\overline{AB})$ or $m(\overline{BA})$. This measure, as in Chapter 5, is always a positive number. With this understanding we see that the measure of the distance from A to B in Illus. 8.2 is 6 units, and the measure of the distance from B to A is also 6 units. This distance measure has several properties that should be fairly obvious. First, the measure of the distance from A to B is equal to the measure

230

of the distance from B to A. Second, the measure of the distance from A to A is 0. Third, the measure of the distance from A to B plus the measure of the distance from B to C is always greater than or equal to the measure of the distance from A to C. Symbolically we can summarize these properties as

$$m(\overline{AB}) = m(\overline{BA})$$
$$m(\overline{AA}) = 0$$
$$m(AB) + m(\overline{BC}) \geq m(\overline{AC})$$

We turn our attention once again to the window and the spider. Illustration 8.12 shows that the spider has moved from his original position to a new one. How great a distance has he covered, if we assume he

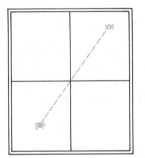

ILLUSTRATION 8.12   The spider moves to a new location

travelled in a straight line? The formulas developed thus far have told us how to find the measure of a distance between two points on a vertical or horizontal line, but tell us nothing about finding the measure of a distance along a diagonal. It is not necessary to consider diagonal measure as a special case, for we can express the distance from the spider's original resting spot to his new one in terms of distances along the vertical and horizontal axes. To help us in this situation, we have drawn a rectangular coordinate system in Illus. 8.13 and have located two points $R(x_1, y_1)$ and $S(x_2, y_2)$ on it. This figure shows both R and S in the first quadrant, with the line segment $\overline{RS}$ inclined to the x axis. We shall use this position of R and S in the development of the formula. Since $m_d(\overline{OA_1}) = x_1$ and $m_d(\overline{OA_2}) = x_2$, the distance from A_1 to A_2 is $x_2 - x_1$. Similarly, the distance from B_1 to B_2 is $y_2 - y_1$. We know then that $m_d(\overline{RZ}) = x_2 - x_1$ and that $m_d(\overline{ZS}) = y_2 - y_1$. (What property developed earlier supports these two statements?) Triangle RZS is a right triangle, and we may use

231

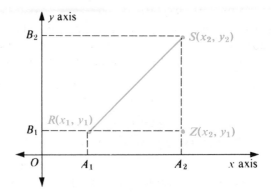

ILLUSTRATION 8.13   The distance between two points in the plane

the Pythagorean theorem:

$$m(\overline{RS})^2 = (x_2 - x_1)^2 + (y_2 - y_1)^2$$

and $$m(\overline{RS}) = \pm \sqrt{(x_2 - x_1)^2 + (y_2 - y_1)^2}$$

Since we are not concerned about direction, the measure assigned to the distance from R to S will be positive; hence,

Formula 8.1 The distance between two points $R(x_1, y_1)$ and $S(x_2, y_2)$ in the plane is

$$m(\overline{RS}) = + \sqrt{(x_2 - x_1)^2 + (y_2 - y_1)^2}$$

In our original considerations x_1, x_2, y_1, and y_2 were all positive, with $x_2 > x_1$ and $y_2 > y_1$.   However, this is not vital to our derivation. Since either point can be called R or S, if the order of the two points is changed, then we have $(x_1 - x_2)^2$ instead of $(x_2 - x_1)^2$. But since

$$(x_2 - x_1)^2 = (x_1 - x_2)^2$$

the order of subtraction is unimportant. The formula for $m(\overline{RS})$ holds for any two points R and S.

To see how this formula is used, consider the two points in Illus. 8.14. R has coordinates $(1, 3)$ and S has coordinates $(4, 6)$. Using Formula 8.1 for the distance between the two points, we find

$$m(\overline{RS}) = \sqrt{(x_2 - x_1)^2 + (y_2 - y_1)^2}$$
$$m(\overline{RS}) = \sqrt{(4 - 1)^2 + (6 - 3)^2}$$
$$m(\overline{RS}) = \sqrt{3^2 + 3^2} = \sqrt{18}$$
$$m(\overline{RS}) = 3\sqrt{2}$$

232

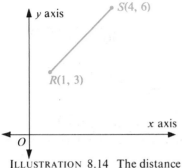

ILLUSTRATION 8.14 The distance from $R(1, 3)$ to $S(4, 6)$ is $3\sqrt{2}$

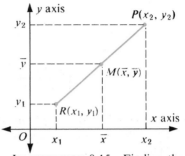

ILLUSTRATION 8.15 Finding the midpoint of $\overline{RS}$

Similarly, the distance from S to R is

$$m(\overline{SR}) = \sqrt{(1 - 4)^2 + (3 - 6)^2}$$
$$m(\overline{SR}) = \sqrt{18} = 3\sqrt{2}$$

Let us suppose that the original spot occupied by the spider had coordinates $(7, 3)$ and his new location coordinates $(-5, -1)$. We can find how far he travelled by using Formula 8.1; the distance he traveled is

$$\sqrt{[7 - (-5)]^2 + [3 - (-1)]^2} = \sqrt{12^2 + 4^2} = \sqrt{169} = 13 \text{ units}$$

The distance formula can be used to help us develop more analytic geometry. In Illus. 8.15 we have shown $R(x_1, y_1)$ and $P(x_2, y_2)$ and the line segment connecting them. The point M in the diagram is the midpoint of this segment. How can we find the coordinates of M? Suppose we assign M the coordinates $(\bar{x}, \bar{y})$. Since $\bar{x}$ is one-half of the way between x_1 and x_2, we have $x_2 - \bar{x} = \bar{x} - x_1$. (Do you see why this is true?) If we solve this equation for $\bar{x}$ we get

$$x_2 - \bar{x} = \bar{x} - x_1$$
$$2\bar{x} = x_2 + x_1$$
$$\bar{x} = \tfrac{1}{2}(x_2 + x_1)$$

Using a similar argument, the formula for $\bar{y}$ in terms of y_1 and y_2 is $\bar{y} = \tfrac{1}{2}(y_2 + y_1)$. We can verify that these are the coordinates of the midpoint of the line joining R and P by using the distance formula to show $m(\overline{RM}) = m(\overline{MP})$. Thus

$$m(\overline{RM}) = \sqrt{[x_1 - \tfrac{1}{2}(x_2 + x_1)]^2 + [y_1 - \tfrac{1}{2}(y_2 + y_1)]^2}$$
$$= \sqrt{\frac{x_2{}^2 - 2x_1x_2 + x_1{}^2 + y_2{}^2 - 2y_1y_2 + y_1{}^2}{4}}$$

233

and similarly

$$m(\overline{MP}) = \sqrt{[x_2 - \tfrac{1}{2}(x_1 + x_2)]^2 + [y_2 - \tfrac{1}{2}(y_2 + y_1)]^2}$$

$$= \sqrt{\frac{x_2^2 - 2x_1x_2 + x_1^2 + y_2^2 - 2y_1y_2 + y_1^2}{4}}$$

Since $\overline{RM}$ and $\overline{MP}$ have equal measure, M is the midpoint of $\overline{RP}$.

The midpoint of the line segment joining the points $Q(-3, 5)$ and $R(3, 1)$ has coordinates

$$\bar{x} = \tfrac{1}{2}(x_2 + x_1) = \tfrac{1}{2}(3 - 3) = 0$$

$$\bar{y} = \tfrac{1}{2}(y_2 + y_1) = \tfrac{1}{2}(5 + 1) = 3$$

Hence, the midpoint of the line segment joining Q and R has coordinates $(0, 3)$.

EXPERIMENT 8B

1 Locate the points $A(2, 3)$ and $B(4, 7)$. Draw the line segment connecting them. How great is the change in the y coordinate from A to B? How great is the change in the x coordinate from A to B? How would you describe the position of the segment relative to the axes?

2 Locate the points $A(2, 3)$ and $B(4, 11)$. Draw the line through them. Answer the questions asked in Experiment 8B(1). How does this line differ from the one drawn in Experiment 8B(1)?

3 Locate the points $A(2, 3)$ and $B(4, 5)$. Draw the line connecting them. Answer the questions asked in Experiment 8B(1). How does this line differ from the one drawn in Experiment 8B(1)?

4 Locate the points $A(2, 3)$ and $B(4, 3)$. Draw the line connecting them. How would you describe the position of this line relative to the axes? How much does x change from A to B? How much does y change from A to B?

5 Locate $A(4, 3)$ and $B(0, 1)$. Draw the line connecting them. How would you describe the position of this line relative to the axes? Is it different from the descriptions in Experiments 8B(1), 8B(2), and 8B(3)?

6 Find two points on the number line which are 2 units apart. Find another two points which are 2 units apart. Find several pairs of points 2 units apart on the positive x axis. Can you write a rule which relates the coordinates assigned to any of these pairs of points to the number 2?

7 Construct the coordinate plane and locate ten points at random. How do you decide whether any three are on the same straight line?

234

EXERCISE 8B

1 Find the distance between each set of points:
   (a) $A(0, 7)$ and $B(5, 3)$ (b) $A(-1, 4)$ and $B(4, -1)$
   (c) $A(5, 0)$ and $B(-5, 4)$ (d) $A(6, -2)$ and $B(-5, 4)$

2 Which of the following sets of coordinates refer to points on a straight line? Plot each set, then use analytic means to confirm your answer.
   (a) $A(1, 2), B(3, 6), C(-2, 4)$ (b) $A(0, 0), B(2, 8), C(8, 2)$
   (c) $A(6, -5), B(2, 1), C(4, -2)$ (d) $A(1, 1), B(1, 2), C(1, 5)$

3 Find the midpoint of the line segments joining
   (a) $A(0, 7)$ and $B(5, 3)$ (b) $A(-1, 4)$ and $B(4, -1)$
   (c) $A(5, 0)$ and $B(-5, 4)$ (d) $A(6, -2)$ and $B(-5, 4)$

4 A triangle has vertices at $(0, 0)$, and $(4, 0)$, and $(2, 4)$. Show that this triangle has two equal sides.

5 A triangle has vertices at $(1, 2), (5, 2),$ and $(5, 6)$. Is it a right triangle?

6 If three vertices of a parallelogram are $(1, -3), (6, 5),$ and $(4, 8)$, find a possible fourth vertex. Are there more than one?

7 Plot $(3, 5), (5, -3), (-3, 5),$ and $(-5, 3)$. Can a rectangle be constructed using these points as vertices?

8 Find the lengths of the sides of a triangle with vertices at $(5, 0), (8, 2),$ and $(6, 5)$. What kind of triangle is this?

9 Show that the points $(5, 0), (8, 2), (6, 5),$ and $(3, 3)$ are the vertices of a square.

10 What is the perimeter of the quadrilateral with vertices at $(-1, 3), (2, 2), (2, -4),$ and $(-2, -3)$?

11 Plot the points $(4, 0), (3, 1), (2, 2), (1, 3),$ and $(0, 4)$ on the same set of axes. What do you notice about the location of these points? If the first number of the pairs is represented by x and the second number of the pair is represented by y, what can be said about $x + y$ for each pair?

12 Draw the line through the points $(3, 1)$ and $(1, 3)$. What other points are on this line? Can you write a number sentence for this line; that is, can you write a formula to relate each x to a y?

13 Draw the line through $(3, 1)$ and $(7, 1)$. What other points are on this line? Can you write a rule for the y value of this line?

14 You are told a line goes through the origin and the point $(3, 3)$. Name two other points it passes through. How can you relate each x to a y on this line?

15 A line passes through $(0, 2)$ and $(3, 5)$. Does it also pass through $(-2, 0)$? How would you relate x and y for points on this line?

235

8.3   SLOPE OF A LINE

The experiments in the last section should make it clear that not all lines in the plane extend in the same direction. In those experiments, some of the lines went from lower left to upper right, and others from upper left to lower right. The "slant" of a line is called its *slope*. The slope of a line is a help in analyzing the line in the plane. As a beginning, we state that a horizontal line has a slope of 0. Now consider the line in Illus. 8.16.

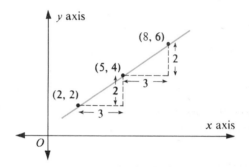

ILLUSTRATION 8.16   The slope of a line is the ratio $(y_2 - y_1)/(x_2 - x_1)$

Notice that it slopes upward from left to right, gaining an altitude of 2 units for each 3 units of run to the right. The slope of a line is the ratio (change in y)/(change in x) per unit along the line. In this example, the slope would be $\frac{2}{3}$. Notice the slope of the line is $\frac{2}{3}$ along the entire length of the line. There are 6 units of altitude gained in 9 units of run. That is, the slope is $\frac{6}{9} = \frac{2}{3}$. More formally, we can state the definition for slope as

Definition 8.1   Given a line through two points A and B with coordinates (x_1, y_1) and (x_2, y_2), respectively; the slope m of the line is the ratio $(y_2 - y_1)/(x_2 - x_1)$.

The line shown in Illus. 8.16 has positive slope. The slope of a line may be positive, negative, zero, or undefined. We will see in the exercises when each of these cases occurs. According to the convention of the definition, a line with positive slope rises to the right while a line with negative slope rises to the left. If $y_1 = y_2$ then $m = 0$ and the line is horizontal. This is consistent with our earlier statement that the slope of a horizontal line should be zero. If $x_1 = x_2$ the line is vertical and the slope is undefined. Do you see why? Since only vertical lines have undefined slopes, we can recognize this situation when it occurs.

236

To find the slope of the line passing through the points $A(-3, 5)$ and $B(2, 3)$ we use the formula given in the definition:

$$m = \frac{y_2 - y_1}{x_2 - x_1} = \frac{3 - 5}{2 - (-3)} = \frac{-2}{5}$$

Since the value of m is negative, we know that the line rises to the left (Illus. 8.17).

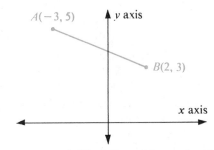

ILLUSTRATION 8.17 A line with negative slope

EXPERIMENT 8C

1 The three points $A(1, -1)$, $B(2, 1)$, and $C(3, 3)$ are on a straight line. What is the slope from A to B? What is the slope from B to C?

2 For three points A, B, and C, the slope from A to B is $\frac{1}{2}$, and the slope from B to C is $\frac{1}{3}$. Can these three points be on a straight line?

3 Plot the points $A(2, 1)$, $B(0, 0)$, $C(6, 4)$, and $D(4, 3)$. Draw the lines $\overleftrightarrow{AC}$ and $\overleftrightarrow{BD}$. Determine the slope of each of these lines.

4 Plot the points $A(4, 3)$, $B(6, 5)$, $C(-2, 3)$, and $D(0, 5)$. Draw the lines $\overleftrightarrow{AC}$ and $\overleftrightarrow{BD}$. Determine the slope of each of these lines. Are the results of this experiment consistent with the results in Experiment 8C(1) above?

5 Locate the points $A(-2, 4)$, $B(-2, 2)$, $C(0, 2)$, and $D(0, 0)$. Draw the lines $\overleftrightarrow{AC}$ and $\overleftrightarrow{BD}$. Determine the slope of $\overleftrightarrow{AC}$. Without calculating, what would you guess the slope of $\overleftrightarrow{BD}$ to be? Check your guess by calculating the slope.

6 Locate the points $A(-2, 2)$, $B(2, 2)$, $C(2, -2)$, and $D(-2, -2)$. Draw the lines $\overleftrightarrow{AC}$ and $\overleftrightarrow{BD}$. Determine the slopes of the lines. Are they equal? What is the product of the two slopes?

7 Locate the points $A(2, 1)$, $B(0, 3)$, $C(3, 3)$, and $D(2, 0)$. Draw the lines $\overleftrightarrow{AC}$ and $\overleftrightarrow{BD}$. Determine the slopes of the lines. Are they equal? What is the product of the slopes?

8 Draw a line with a slope of $\frac{1}{2}$. Can you draw another line with a slope of $\frac{1}{2}$?

237

9 Draw a line with a slope of 2 that goes through the point (1, 3). Can you draw another distinct line through (1, 3) which also has a slope of 2?

10 Draw a line with slope of $\frac{1}{4}$ through the point (−1, 0). Can you draw a line with slope $\frac{1}{4}$ that goes through both (−1, 0) and (2, 3)?

11 Consider a line through (2, 0) and (2, 4). What are the difficulties in writing the slope of this line?

EXERCISE 8C

1 Given the following sets of points, determine which are on straight lines.
   (a) $A(3, 1), B(5, 8), C(8, 15)$ (b) $A(-5, 2), B(0, 7), C(3, 10)$
   (c) $A(3, 9), B(2, 8), C(1, 7)$ (d) $A(8, 9), B(17, 2), C(-1, 16)$

2 Calculate the slope of a line through each pair of points:
   (a) $A(2, 2)$ and $B(4, 1)$ (b) $A(-1, -3)$ and $B(-5, -8)$
   (c) $A(3 + b, 4 + b)$ and $B(5 + b, 2 + b)$ (d) $A(3k, 4k)$ and $B(3k + 1, 4k + 1)$

3 Let A, B, C, and D represent the vertices of a quadrilateral. Determine which of the following are parallelograms.
   (a) $A(4, 4), B(0, 0), C(5, 0), D(5, 4)$
   (b) $A(-2, 1), B(3, 1), C(1, -1), D(-4, -1)$
   (c) $A(3, -5), B(4, -6), C(-1, 6), D(-2, -5)$
   (d) $A(8, 2), B(-1, 3), C(-2, -6), D(5, -4)$

4 What is the slope of the line through (1, 2) and (3, 5)? How does it relate to the slope of the line through (2, 4) and (6, 10)?

5 What is the slope of the line through (2, −3) and (1, 1)? How does it relate to the slope of the line through (10, −15) and (5, 5)? Generalize from your answers to Exercises 8C(4) and 8C(5).

6 The diagram shows the side view of the roof of a house. Carpenters describe the slope of the rafters by forming a ratio of rise/run. Is this consistent with our definition of slope?

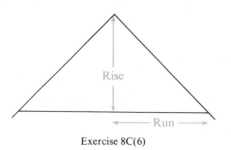

Exercise 8C(6)

7 Find the slopes of the non-parallel sides of the isosceles trapezoid with vertices at $A(-3, 8), B(3, 8), C(-6, 3), D(6, 3)$.

238

8.4 PARALLELS AND PERPENDICULARS

The last set of experiments called for several pairs of lines with the same slope. If we plot the points $A(1, 1)$, $B(2, 1)$, $C(5, 4)$, $D(6, 4)$, and draw $\overleftrightarrow{AC}$ and $\overleftrightarrow{BD}$ (Illus. 8.18), it appears the lines are parallel. Since the slope of $\overleftrightarrow{AC}$ is $(4 - 1)/(5 - 1) = \frac{3}{4}$ and the slope of $\overleftrightarrow{BD}$ is $(4 - 1)/(6 - 2) = \frac{3}{4}$, these two lines have the same slope. This result and your work in the ex-

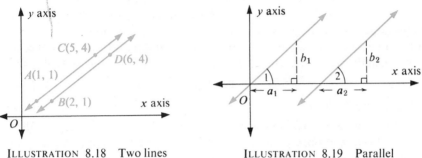

ILLUSTRATION 8.18 Two lines
with the same slope

ILLUSTRATION 8.19 Parallel
lines have the same slope

periments should lead you to surmise that the slopes of two parallel lines are always equal. Remember, however, that these examples are not enough to establish the conjecture. In this case, we can show that the conjecture is true. Consider the parallel lines in Illus. 8.19. If line 1 is parallel to line 2, what is true of $\measuredangle 1$ and $\measuredangle 2$? The segments with measures b_1 and b_2 are drawn perpendicular to the x axis. Do we know that the two triangles are similar? How? If $a_1 = a_2$, then what do we know about the triangles? Does it follow that $b_1/a_1 = b_2/a_2$? Does this lead you to believe that parallel lines always have the same slope? In Illus. 8.20 we have

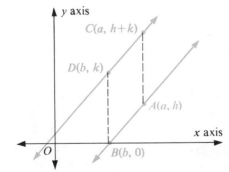

ILLUSTRATION 8.20 Finding the slope of parallel lines

239

used an analytic approach to establish the conjecture. An axis has been superimposed on two parallel lines so that one line crosses the x axis at B. You will notice that the point D directly above B is given the coordinates (b, k). It is easy to see that the x coordinate of D is b. By stating that the y coordinate is k, we are saying that if a line is dropped from D to B, perpendicular to the x axis, the measure of the distance from D to B is k units. If we establish that A has coordinates (a, h), then do you see that if the lines in Illus. 8.20 are parallel, C must have the coordinates (a, h + k)? The slope of $\overleftrightarrow{AB}$ is h/(a − b), and the slope of $\overleftrightarrow{CD}$ is (h + k − k)/(a − b) = h/(a − b), and these two parallel lines have the same slope. Since all the statements we made were general, that is, they did not depend upon a particular position (excluding vertical) or slope, we can conclude that non-vertical parallel lines always have the same slope.

THEOREM 8.2 *Two distinct non-vertical lines are parallel if and only if their slopes are the same.*

The last set of experiments might also lead one to suspect that there is some fixed relationship between the slopes of perpendicular lines. Two perpendicular lines are shown in Illus. 8.21; they were drawn by passing both lines through the point (0, 0) and one each through the points (− 3, 3) and (3, 3). You can verify by using the Pythagorean theorem that the angle at the origin is indeed a right angle and consequently the two lines $\overleftrightarrow{OA}$ and $\overleftrightarrow{OB}$ are perpendicular. The slope of $\overleftrightarrow{OA}$ is +1 and the slope of $\overleftrightarrow{OB}$ is − 1, and the product of the two slopes is − 1. Does this agree with your work in the experiments? This gives us some insight into the slopes of perpendicular lines but, before we draw any conclusions, we need to investigate a more general case.

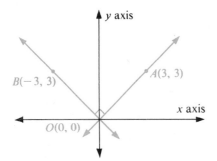

ILLUSTRATION 8.21 Perpendicular lines

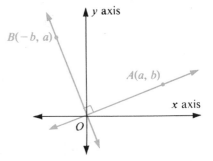

ILLUSTRATION 8.22 The product of the slopes of perpendicular lines is − 1

240

Consider now Illus. 8.22. It shows the coordinate axes superimposed on a set of perpendicular lines so that the vertex of the right angle is at $(0, 0)$. Although it is convenient for us to study the perpendicular lines located in this position, there is no loss in generality in the discussion if the axes are located elsewhere. A point has been selected on $\overleftrightarrow{OA}$ and marked with the coordinates (a, b). By use of the Pythagorean relationship, we can determine that there is a point on $\overleftrightarrow{OB}$ with coordinates $(-b, a)$. The slope of $\overleftrightarrow{OA}$ can be calculated as

$$\frac{b - 0}{a - 0} = \frac{b}{a}$$

and the slope of $\overleftrightarrow{OB}$ is

$$\frac{0 - a}{0 - (-b)} = -\frac{a}{b}$$

You will notice that the slopes of these two lines are negative reciprocals of each other, and that the product of their slope is

$$\left(\frac{b}{a}\right)\left(-\frac{a}{b}\right) = -1$$

Hence, we confirm that perpendicular lines always have slopes which are negative reciprocals of each other.

THEOREM 8.3   *If two lines are perpendicular, then the product of their slopes is* -1.

EXPERIMENT 8D

1 Plot each of the following pairs of points in the same coordinate system. Draw $\overleftrightarrow{AB}$ in each case and find its slope.
(a) $A(0, 0)$, $B(5, 0)$ (b) $A(0, 0)$, $B(5, 1)$
(c) $A(0, 0)$, $B(5, 3)$ (d) $A(0, 0)$, $B(5, 5)$
(e) $A(0, 0)$, $B(3, 5)$ (f) $A(0, 0)$, $B(1, 5)$
(g) $A(0, 0)$, $B(-1, 5)$ (h) $A(0, 0)$, $B(-3, 5)$

2 Plot each of the following pairs of points in the same coordinate system. Draw $\overleftrightarrow{AB}$ in each case and find its slope.
(a) $A(-2, -3)$, $B(0, 0)$ (b) $A(0, -3)$, $B(2, 0)$
(c) $A(-2, 0)$, $B(0, 3)$ (d) $A(0, 0)$, $B(2, 3)$

3 On a single coordinate system, draw lines which pass through the origin and have
(a) A slope of 0   (b) A slope of 1   (c) A slope of 2
(d) A slope of 10   (e) A slope of -5 (f) A slope of -1

241

4 Plot $A(0, 0)$, $B(0, 6)$, and $C(5, 6)$. Draw line segments $\overline{AB}$, $\overline{AC}$, and $\overline{BC}$. Find the area of $\triangle ABC$.

5 Plot $A(-3, 0)$, $B(3, 0)$, and $C(0, 4)$. Draw line segments $\overline{AB}$, $\overline{AC}$, and $\overline{BC}$. Find the area of $\triangle ABC$.

6 Plot $A(1, 1)$, $B(6, 1)$, and $C(3, 5)$. Draw line segments $\overline{AB}$, $\overline{AC}$, and $\overline{BC}$. Find the area of $\triangle ABC$.

7 Plot $A(3, 2)$, $B(7, 2)$, and $C(-3, 5)$. Draw line segments $\overline{AB}$, $\overline{AC}$, and $\overline{BC}$. Find the area of $\triangle ABC$.

8 A quadrilateral has vertices at $(1, 1)$, $(5, 1)$, $(2, 4)$, and $(4, 4)$. Show that it is an isosceles trapezoid.

9 What is the slope of a line perpendicular to a horizontal line?

EXERCISE 8D

1 Which of the following pairs of points are on lines which are parallel?
   (a) $A(1, 6)$, $B(2, 5)$ and $C(3, -2)$, $D(6, 2)$
   (b) $A(0, 3)$, $B(6, 0)$ and $C(1, 0)$, $D(-1, 2)$
   (c) $A(-1, 0)$, $B(1, 0)$ and $C(-5, 0)$, $D(5, 0)$
   (d) $A(2, 2)$, $B(4, 0)$ and $C(4, 2)$, $D(2, 4)$

2 Which of the following pairs of points are on lines which are perpendicular?
   (a) $A(4, 0)$, $B(5, 1)$ and $C(2, 2)$, $D(0, 4)$
   (b) $A(6, 2)$, $B(7, -1)$ and $C(6, 8)$, $D(2, -4)$
   (c) $A(-1, -1)$, $B(1, 1)$ and $C(-1, 1)$, $D(1, -1)$
   (d) $A(4, 1)$, $B(6, -1)$ and $C(-2, 5)$, $D(0, 3)$

3 Why is the slope of a line undefined when two of its points have the same x coordinate?

4 Would we have been able to complete the proof of Theorem 8.3 if the lines had been positioned parallel to the axes? Does the proof hold for lines in this position?

5 Show that the Pythagorean relationship holds in Illus. 8.21.

6 In each case show that the lines determined by the sets of points are not perpendicular.
   (a) $A(5, 2)$, $B(3, 1)$ and $C(2, 4)$, $D(-1, -5)$
   (b) $A(6, 3)$, $B(2, 8)$ and $C(-1, 6)$, $D(4, 4)$

7 If two lines are perpendicular and one has a slope of $-\frac{1}{3}$, what is the slope of the other?

8 One side of a square has a slope of 2. What are the slopes of the other sides?

9 Can a line have a slope of 100? Of 1000? Of -0.0001?

10 Can a triangle have three sides with slopes of $\frac{1}{2}$, 2, and 5? Why?

242

8.5 THE AREA OF A TRIANGLE

In the last set of experiments, you were asked to find the areas of several triangles. In the first, a right triangle, it was easy to determine the base and the height and find the area using the formula $A = \frac{1}{2}bh$. In the second, it was also easy to determine the base and height and to use the formula to find the area. After that the problem became much more complicated. In each case, to use the formula it was necessary to find the altitude. Although this is simple to do in some triangles, in others it is quite difficult. In each case the triangle was described by the three vertex points. Illustration 8.23 shows a triangle determined by the points A, B,

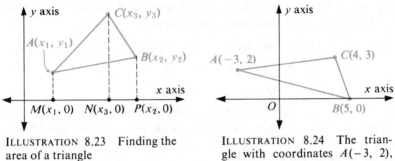

ILLUSTRATION 8.23 Finding the area of a triangle

ILLUSTRATION 8.24 The triangle with coordinates $A(-3, 2)$, $B(5, 0)$, and $C(4, 3)$ has an area of 11

and C with coordinates (x_1, y_1), (x_2, y_2), and (x_3, y_3), respectively. We shall use it to find a formula for the area of a triangle whose coordinates are given, as in Illus. 8.24. The area of triangle ABC can be determined by considering the larger figure $ACBPM$, since it is equal to the area of the trapezoid $MACN$ plus the area of trapezoid $NCBP$ minus the area of trapezoid $MABP$. From Chapter 5 we know the area of a trapezoid is equal to one-half the sum of the measures of its bases times the measure of its altitude. Therefore, using the distance formulas we can restate the relation

Area (ABC) = area $(MACN)$ + area $(NCBP)$ − area $(MABP)$

as

$$\text{Area} = \frac{1}{2}(y_1 + y_3)(x_3 - x_1) + \frac{1}{2}(y_2 + y_3)(x_2 - x_3)$$
$$- \frac{1}{2}(y_1 + y_2)(x_2 - x_1)$$
$$= \frac{1}{2}(y_1 x_3 - x_1 y_1 + y_3 x_3 - y_3 x_1 + y_2 x_2 - y_2 x_3 + x_2 y_3$$
$$- x_3 y_3 - y_1 x_2 + y_1 x_1 - y_2 x_2 + x_1 y_2)$$
$$= \frac{1}{2}(x_1 y_2 + x_2 y_3 + x_3 y_1 - x_2 y_1 - x_3 y_2 - x_1 y_3).$$

243

This formula, although cumbersome to look at, gives the area of a triangle in terms of the coordinates of the vertices. The order of the labelling of the vertices is unimportant, but once the values are assigned, they must be kept straight in the formula. We can use this formula to find the area of the triangle in Illus. 8.24 with coordinates $A(-3, 2)$, $B(5, 0)$, and $C(4, 3)$:

$$\text{Area} = \tfrac{1}{2}[(-3)(0) + (5)(3) + (4)(2) - (5)(-2) - (4)(0) - (-3)(3)]$$
$$= \tfrac{1}{2}(0 + 15 + 8 - 10 - 0 + 9)$$
$$= \tfrac{1}{2}(22) = 11$$

Suppose the points had been named as $A(-3, 2)$, $B(4, 3)$, and $C(5, 0)$. Surely this is the same triangle described before and should have the same area. But if we substitute in the formula we find

$$\text{Area} = \tfrac{1}{2}[(-3)(3) + (4)(0) + (5)(2) - (4)(2) - (5)(3) - (-3)(0)]$$
$$= \tfrac{1}{2}(-9 + 0 + 10 - 8 - 15 - 0)$$
$$= \tfrac{1}{2}(-22) = -11$$

You can see that this is the negative of the first result. Since area cannot be negative, it would seem reasonable to assume that we should simply ignore a negative sign if it appears. However, just "dropping" the sign is bound to make even a careless student a little upset unless we can find some reason for its appearance. Notice in the discussion in which the formula was developed that the points A, B, and C appear in a counterclockwise pattern. In our first example the points named also appeared in a counterclockwise pattern, and the result was positive. In the second the points did not appear in a counterclockwise pattern, and so we had a discrepancy in the sign. What we found was a *directed area*, related to the concept of directed distance. Recall that we were interested in the distance between two points without regard to sign. Similarly we look for a positive value for area, rather than a directed area.

EXPERIMENT 8E

1 Derive a formula for the area of a triangle which can be used for the special cases when one of its vertices is the point $(0, 0)$.

2 Derive a special formula to use for the area of a triangle when the coordinates of its vertices are $A(x_1, y_1)$, $B(x_2, y_1)$, and $C(x_2, y_2)$. What kind of triangle is described by these vertices?

3 Construct a triangle with one side on the x axis. Find the midpoint of each side of the triangle and draw a line segment connecting the midpoints. Measure this segment and the base. How do they compare? Repeat for

several more triangles. Do you see a relationship between the length of the segment joining the midpoints and the length of the base?

4 Construct a parallelogram $ABCD$. Draw the diagonals and label the point of intersection X. Measure $\overline{AX}$ and $\overline{XC}$. Do they have the same measure? Measure $\overline{BX}$ and $\overline{XD}$. Do they have the same measure? Repeat this experiment for several parallelograms. What relationships appear to exist?

5 Construct a rectangle. Draw and measure the diagonals. Are they the same length? Repeat for several other rectangles. Does each experiment reinforce your first conclusion?

6 Draw a quadrilateral $ABCD$. Find the midpoint of each side. Connect these points in order. What new figure did you draw? Repeat for several other quadrilaterals. Is the result always the same?

7 Construct an equilateral triangle. Find the midpoint of each side. Connect these points in order. What new figure did you draw? Does this appear to be another equilateral triangle?

8 Construct a rhombus. Measure the diagonals. Are they equal in measure? Repeat several times. What conclusion would you draw?

9 Construct a general triangle. Find the midpoint of each side. Connect these points in order. Does the new figure appear to be similar to the original triangle? Repeat for several other triangles.

10 Draw a circle. Draw two diameters; label the endpoints of the first A and B, and the endpoints of the second C and D. Draw $\overline{AC}$ and $\overline{BD}$. Is the measure of $\overline{AC}$ the same as the measure of $\overline{BD}$? Repeat two more times. What conclusion can you draw?

EXERCISE 8E

1 Using the diagram, derive a formula for the area similar to that derived in the text.

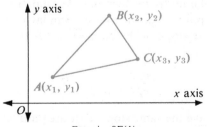

Exercise 8E(1)

2 Find the area of the triangle with vertices at $A(3, 5)$, $B(8, 2)$, and $C(5, 5)$.

3 Find the area of the triangle with vertices at $A(-6, -2)$, $B(5, 4)$, and $C(2, -7)$.

245

4 Using the formula derived in Experiment 8E(1), find the area of the triangle with vertices at $A(0, 0)$, $B(2, 3)$, and $C(4, 8)$.

5 Using the formula derived in Experiment 8E(2), find the area of the triangle with vertices at $A(3, 4)$, $B(-6, 4)$, and $C(-6, 5)$.

6 Find the area of the quadrilateral whose vertices are $A(3, 1)$, $B(7, 2)$, $C(6, 6)$, and $D(-2, 4)$.

7 Find the area of the quadrilateral whose vertices are $A(-2, -5)$, $B(6, 3)$, $C(8, -1)$, and $D(-3, -3)$.

8 Develop a formula for the area of a square using the procedure in this section.

9 Develop a formula for the area of a rectangle.

10 Develop a formula for the area of a parallelogram.

8.6 USING ANALYTIC GEOMETRY IN PROOFS

The last set of experiments should have led you to some intuitive conclusions. Let us investigate now how we might prove these conjectures using what we have learned in analytic geometry. For one, you might have concluded that

> *The line joining the midpoints of the sides of a triangle is parallel to the base and has measure equal to one-half the measure of the base.*

Construct a triangle on a coordinate axis (Illus. 8.25). There is no loss in generality if the base rests on the x axis and if one vertex is on the y axis. Positioning the triangle in this manner will, however, make it easier to follow the proof. We must be careful not to allow the proof to depend upon the triangle being a special kind of triangle, but to have a proof which is independent of any special characteristics. Illustration 8.25 is not a right or equilateral triangle, nor does it have any other special properties. Using the midpoint formula, we find the midpoint of $\overline{AB}$ is $(a/2, b/2)$. This is point D in the figure. Similarly, point E has coordinates $(c/2, b/2)$. The slope of the base $\overline{AC}$ is $(0-0)/(c-a) = 0$, and the slope of $\overline{DE}$ is

$$\frac{b/2 - b/2}{c/2 - a/2} = 0$$

Since $\overline{AC}$ and $\overline{DE}$ have the same slope, they are parallel. The length of the base, $\overline{AC}$, is $\sqrt{(c-a)^2}$ and the length of $\overline{DE}$ is

$$\sqrt{\left(\frac{c}{2} - \frac{a}{2}\right)^2 + \left(\frac{b}{2} - \frac{b}{2}\right)^2} = \tfrac{1}{2}\sqrt{(c - a)^2}$$

246

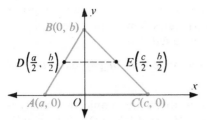

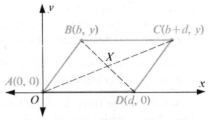

ILLUSTRATION 8.25 The line joining the midpoints of the sides of a triangle is parallel to the base and has measure equal to one-half the measure of the base

ILLUSTRATION 8.26 The diagonals of a parallelogram bisect each other

Hence, the length of $\overline{DE}$ is one-half the length of the base $\overline{AC}$. Using analytic techniques, it has been easy for us to establish that the line joining the midpoint of two sides of a triangle is parallel to the base and has measure equal to one-half the measure of the base. A comparable deductive proof without the use of analytical techniques is much longer.

Experiment 8E(4) might have led you to the conclusion that

The diagonals of a parallelogram bisect each other.

To establish this conjecture analytically, we construct a parallelogram on a coordinate axis as in Illus. 8.26. The parallelogram has been positioned with one vertex at $(0, 0)$ and with one side along the x axis. Do you think the generality of our proof will be hurt by this positioning? Why?

The midpoint of diagonal $\overline{AC}$ is $((b+d)/2, y/2)$, and the midpoint of diagonal $\overline{BD}$ is $((b+d)/2, y/2)$. Since the midpoint of a line segment must also be on the line segment, the point $((b+d)/2, y/2)$ must be on both $\overline{AC}$ and $\overline{BD}$. In the illustration the one point $\overline{AC}$ and $\overline{BD}$ have in common has been labeled X. It is the point with coordinates $((b + d)/2, y/2)$. Notice now that the length of $\overline{AX}$ is

$$\sqrt{\left(\frac{b + d}{2}\right)^2 + \left(\frac{y}{2}\right)^2}$$

and the length of $\overline{XC}$ is also

$$\sqrt{\left(\frac{b + d}{2}\right)^2 + \left(\frac{y}{2}\right)^2}$$

Hence, $m(\overline{AX}) = m(\overline{XC})$. Similar calculations would show $m(\overline{BX}) = m(\overline{XD})$. Again we have established a conjecture made in the experiments;

247

namely, that the diagonals of a parallelogram bisect each other. A proof of this same conjecture using congruent triangles is more involved.

Perhaps in Experiment 8E(6) you concluded that

Line segments successively joining the midpoints of any quadrilateral form a parallelogram.

We shall use Illus. 8.27 to help us establish this conjecture analytically. The coordinates of E, F, G, and H are $(d/2, z/2)$, $((d + c)/2,$

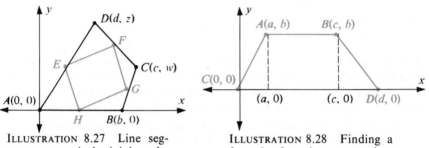

ILLUSTRATION 8.27 Line segments successively joining the midpoints of a quadrilateral

ILLUSTRATION 8.28 Finding a formula for the area of a trapezoid

$(w + z)/2), ((c + b)/2, w/2),$ and $(b/2, 0)$, respectively. The slope of $\overline{EF}$ is

$$\frac{[(w + z)/2] - (z/2)}{[(d + c)/2] - (d/2)} = \frac{w/2}{c/2} = \frac{w}{c}$$

and the slope of $\overline{GH}$ is

$$\frac{w/2}{[(c + b)/2] - (b/2)} = \frac{w/2}{c/2} = \frac{w}{c} .$$

Consequently, $\overline{EF}$ and $\overline{GH}$ are parallel. A similar set of calculations shows that both $\overline{EH}$ and $\overline{FG}$ have a slope $z/(d - b)$. The length of $\overline{EF}$ is

$$\sqrt{\left(\frac{d + c}{2} - \frac{d}{2}\right)^2 + \left(\frac{w + z}{2} - \frac{z}{2}\right)^2} = \tfrac{1}{2}\sqrt{c^2 + w^2}$$

and the length of $\overline{GH}$ is

$$\sqrt{\left(\frac{c + b}{2} - \frac{b}{2}\right)^2 + \left(\frac{w}{2} - 0\right)^2} = \tfrac{1}{2}\sqrt{c^2 + w^2}$$

showing us that not only are $\overline{EF}$ and $\overline{GH}$ parallel, but that they are also equal in length. Similarly, we find that both $\overline{EH}$ and $\overline{FG}$ have a length of $\tfrac{1}{2}\sqrt{(d - b)^2 + z^2}$. This establishes that $EFGH$ is a parallelogram.

248

The same techniques can be used to establish formulas for area. In Illus. 8.28 a trapezoid is shown with one vertex at $(0, 0)$ and with one side along the x axis. The area of the trapezoid is the sum of the areas of the two triangles and the rectangle. The area of a rectangle is found by multiplying the measures of the base and the height; that is,

$$\text{Area}_{\text{Rectangle}} = (c - a)b = cb - ab$$

The area of the left-hand triangle is

$$\text{Area}_{\text{Triangle}} = \tfrac{1}{2}ab$$

and the area of the right-hand triangle is

$$\text{Area}_{\text{Triangle}} = \tfrac{1}{2}(d - c)b = \tfrac{1}{2}bd - \tfrac{1}{2}cb$$

Then the total area is

$$cb - ab + \tfrac{1}{2}ab + \tfrac{1}{2}bd - \tfrac{1}{2}cb = \tfrac{1}{2}cb - \tfrac{1}{2}ab + \tfrac{1}{2}bd$$
$$= \tfrac{1}{2}b(c - a + d)$$

or

$$= \tfrac{1}{2}b[(c - a) + (d - 0)]$$

How does this formula compare with the one we derived in Chapter 5?

EXPERIMENT 8F

1 Using the figure, establish the theorem about bisectors of diagonals in a parallelogram. Do you think it more difficult to establish the theorem using this figure?

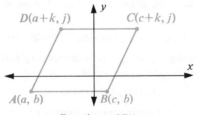

Experiment 8F(1)

2 Construct an isosceles triangle. Find the midpoints of the two equal sides. Draw a line from each midpoint to the vertex opposite it. Measure the two lines you have drawn. What do you conclude? Repeat for several isosceles triangles.

3 Construct a triangle. Construct the perpendicular bisector of each side. Do all three bisectors appear to intersect at one point? Repeat for another triangle.

249

4 Construct a rhombus $ABCD$. Draw its diagonals. Label the point of intersection of the diagonals X. Measure $\overline{AX}$, $\overline{BX}$, $\overline{CX}$, and $\overline{DX}$. Measure each of the angles at X. Do you see a relationship?

5 Construct a right triangle. Find the midpoint of the hypotenuse. Measure the distance from this midpoint to each vertex. Can you generalize?

6 Derive a formula for the area of the trapezoid pictured.

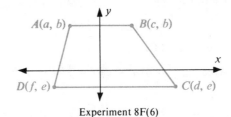

Experiment 8F(6)

7 Draw a line through $(1, 1)$ and $(4, 4)$. On the same axes draw a line through $(2, 1)$ and $(5, 4)$. How would you relate each x and y value on the first line? On the second?

8 Draw a line through $(0, 3)$ and $(3, 9)$. Does this line also pass through $(-2, 1)$. How is each x related to y? Draw a second line through $(0, 5)$ and $(6, 11)$. What relation does it have with the first line? Do they have the same slope?

EXERCISE 8F

In each of the following establish the conjecture analytically.

1 The midpoint of the hypotenuse of a right triangle is equidistant from all three vertices.

2 The diagonals of a rhombus are perpendicular bisectors of each other.

3 The measures of the lines joining the midpoints of the sides of an isosceles triangle to the opposite vertices are equal.

4 The diagonals of a square have equal measure.

5 The figure formed by connecting the midpoints of the sides of an equilateral triangle in order is another equilateral triangle.

6 The measures of the diagonals of an isosceles trapezoid are equal.

7 The sum of the squares of the measures of the sides of a parallelogram is equal to the sum of the squares of the measures of its diagonals.

8 The measures of the diagonals of a rectangle are equal.

9 The altitudes of a triangle intersect in a single point.

10 Prove deductively the property that was shown analytically in Illus. 8.25.

250

8.7 EQUATION OF A STRAIGHT LINE

From work in earlier chapters, we know that two points determine a line. In Section 8.3 we found that the slope of a line is given by the formula $m = (y_2 - y_1)/(x_2 - x_1)$. If we know the slope of a line and one of the points it passes through, do we know enough to draw the line? For the line in Illus. 8.29, suppose we know that $m = \frac{1}{2}$ and that one point on

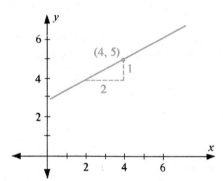

ILLUSTRATION 8.29   The line $2y = x + 6$ passes through $(4, 5)$ and has slope $\frac{1}{2}$

the line, say (x_1, y_1), has coordinates $(4, 5)$. Substituting these quantities in the slope formula gives

$$\frac{1}{2} = \frac{y_2 - 5}{x_2 - 4}$$

or, simplifying, $2y_2 = x_2 + 6$. Since (x_2, y_2) represents any other point on the line, we can write

$$2y = x + 6$$

as the equation of the line.

Formula 8.2   A line with slope m that passes through the point (x_1, y_1) has the equation $y - y_1 = m(x - x_1)$.

This formula is sometimes referred to as the *point-slope equation* for a line. If we examine the equation and Illus. 8.29, we can gain some further information. At what point does the line cut the y axis? It is easy to see the answers to these questions in the illustration, but perhaps not so easy when only the equation is considered.

The point at which a line crosses the y axis is called the y *intercept*;

251

this point has the coordinates $(0, b)$. If we substitute this point in the point-slope equation, we obtain

$$y - y_1 = m(x - x_1)$$
$$y - b = m(x - 0)$$
$$y = mx + b$$

When the equation is written this way, we say it is in *slope-intercept form*. The coefficient of the x term is the slope, and b is the y intercept. Thus, the line with slope $\frac{1}{2}$ that crosses the y axis at the point $(0, 3)$ has the equation $y = \frac{1}{2}x + 3$.

As a second example, suppose a line has a slope of $-\frac{3}{4}$ and passes through the point $(6, -4)$. The equation of the line is given by

$$-\frac{3}{4} = \frac{y + 4}{x - 6}$$

or more simply

$$4y = 3x + 2$$

If this is written in the form

$$y = \tfrac{3}{4}x + \tfrac{1}{2}$$

we know that the line crosses the y axis at $x = 0$, $y = \frac{1}{2}$. A look at Illus. 8.30 will serve to verify these results.

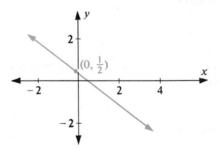

ILLUSTRATION 8.30   The line $y = -\tfrac{3}{4}x + \tfrac{1}{2}$ crosses the y axis at $y = \frac{1}{2}$

EXPERIMENT 8G

1   Draw a triangle and its three altitudes. Does it appear that the altitudes meet in a single point? Repeat for several triangles

2   Draw a triangle and the perpendicular bisectors of the sides of the triangle. Does it appear that the bisectors are concurrent? Repeat for several triangles.

3 All lines parallel to the x axis have a slope of 0. Derive the formulas for all these lines.

4 How do you write the equation of a line parallel to the y axis? What happens in your use of the formulas for lines derived above?

EXERCISE 8G

1 Find the equation of a line that passes through (4, 3) and has slope (a) 2; (b) $\frac{1}{2}$; (c) $-\frac{1}{2}$; (d) -4.

2 Find the equation of the line that passes through (a) (3, 4) and (5, 2); (b) (0, 0) and $(-1, -4)$; (c) (6, 6) and (8, 4).

3 How many distinct lines can you pass through a given point? Through a given point and having a given slope?

4 The vertices of a triangle are (3, 5), $(-2, -6)$, and (1, 2). What are the equations of the medians of the triangles?

5 Prove that the medians of a triangle are concurrent.

6 Prove that the perpendicular bisector of the sides of a triangle are concurrent.

7 Sketch each of the following:
(a) $y = 7x - 5$ (b) $y = -\frac{1}{3}x + 3$ (c) $y = 2x - 4$

8.8 THE EQUATION OF A CIRCLE

Suppose we place a circle on a coordinate system as in Illus. 8.31. The radius of the circle is denoted by r, and the coordinates of the center are (h, k). Consider any point on the circle with coordinates (x, y). Using the

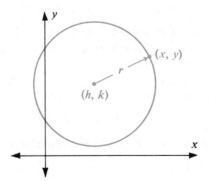

ILLUSTRATION 8.31   The circle with center at (h, k) and radius r has equation $(x - h)^2 + (y - k)^2 = r^2$

253

distance formula we know

$$\sqrt{(x - h)^2 + (y - k)^2} = r$$

or

$$(x - h)^2 + (y - k)^2 = r^2$$

Formula 8.3 The equation of a circle with radius r and with center at (h, k) is given by $(x - h)^2 + (y - k)^2 = r^2$.

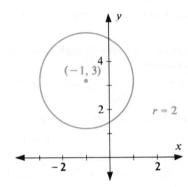

ILLUSTRATION 8.32 The circle with center at $(-1, 3)$ and radius 2 has equation $(x + 1)^2 + (y - 3)^2 = 2^2$

In Illus. 8.32 we have shown a circle with center at $(-1, 3)$ and a radius of 2. The equation of this circle is

$$(x + 1)^2 + (y - 3)^2 = 2^2$$

or, simplifying,

$$x^2 + 2x + y^2 - 6y + 6 = 0$$

This is a somewhat more complicated equation than the kind derived in the last section, and it is a little harder to see if known points on the circle do indeed satisfy this condition.

One of the points on the circle must be $(-1, 1)$. Do you see why? If we substitute these values for x and y, we obtain

$$(-1)^2 + 2(-1) + (1)^2 - 6(1) + 6 = 1 - 2 + 1 - 6 + 6 = 0$$

indicating that this point is on the circle described by the equation.

EXERCISE 8H

1   What is the equation of a circle with a radius of r and center at the origin?

254

2 Write the equation of each of the following circles and sketch its graph.
   (a) Center $(0, 0)$; radius 4   (b) Center $(3, 5)$; radius 1
   (c) Center $(-2, 4)$; radius 6   (d) Center $(-2, -3)$; radius 2

3 Is the point $(0, 0)$ on the circle with center at $(3, 0)$ and radius 3?

4 Is the point $(1, -1)$ on the circle with center at $(2, 3)$ and radius 5?

5 The diameter of a circle has its endpoints at $(-3, 6)$ and $(3, -2)$. What is the equation of the circle?

6 Prove that the angle in the diagram is a right angle.

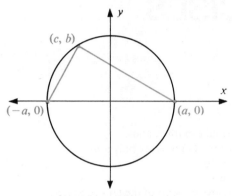

Exercise 8H(6)

7 If a circle has the equation $x^2 + y^2 - 4x - 6y = 12$, what is the equation of a concentric circle with radius 1 unit larger?

8 Write the equation of the circle circumscribed about the triangle whose vertices are $(0, 0)$, $(3, 3)$, and $(-3, 3)$.

9 What is the equation of the circle with center at $(0, 0)$ and $r = \sqrt{2}$?

10 What is the equation of the circle with center at $(-2, -3)$ and $r = a$?

255

ANSWERS TO SELECTED EXERCISES

2   John Reed must take mathematics.
4   All three-sided closed figures are polygons.
6   No valid conclusion
8   No valid conclusion
10 All Nebraskans are residents of the United States.
12 True conclusions are the ones for Exercises 1A(4) and 1A(5); the other valid
    conclusions depend on the individuals mentioned. Yes; yes.

EXERCISE 1B

1   Yes
3   (c), (d) are empty.
5   There is only one letter i, for example, on a keyboard, several different i's in
    a Scrabble game.
7   (a) { }, {$}, {#}, {$, #}
    (b) { }, {line}, {plane}, {space}, {line, plane}, {line, space}, {plane, space},
    {line, plane, space}
9   The set of all people with polka-dot hair; the set of even numbers between $\frac{1}{2}$
    and $\frac{1}{4}$; the set of all circles with five sides; no.
11 Yes, in the first case A and B may have the same number of elements, while
    in the second case we know A contains fewer elements than B.
13 No

256

15

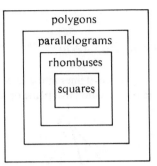

EXERCISE 1C

1 (f), (g) are equal.
3 Yes; not necessarily
5 The elements can be corresponded to the first four counting numbers.
7 (a) No, (b) yes

EXERCISE 1D

1 (a) $\{1, 2, 3, 5\}$; (b) $\{2, 4, 6, 8, 13, 34, 45\}$; (c) $\{1, 2, 3, 4, 5, 6, 7, 8\}$;
 (d) $\{x, y, z\}$; (e) $\{1, 2, 3, x, y, z\}$; (f) $\{1, 4, 9, 16\}$
3 7; A and B are disjoint; 4; $B \subset A$.
5 (a) U; (b) ϕ; (c) U; (d) A; (e) A; (f) ϕ; (g) A; (h) U; (i) U;
 (j) ϕ
7 $\{\ \} \cup \{1, 2, 3\}$; $\{1\} \cup \{2, 3\}$; $\{2\} \cup \{1, 3\}$; $\{3\} \cup \{2, 1\}$
9 When both A and B are empty
11 (a) The set of all rectangles
 (b) The set of all parallelograms
 (c) The set of all non-rectangular parallelograms
 (d) Empty
 (e) The set of all non-rectangular quadrilaterals
 (f) The set of all quadrilaterals that are not parallelograms

EXERCISE 1E

1 (a) Not reflexive, symmetric, not transitive
 (c) Not reflexive, not symmetric, transitive
 (e) Not reflexive, not symmetric, transitive
 (g) Reflexive, symmetric, transitive
3 (a), (f) are linear order relations; (c), (d) are equivalence relations.
5 Not reflexive, not symmetric, not transitive

257

EXERCISE 2A

1 Yes, geometric figure

3 (a) C; (b) $\overrightarrow{CD}$; (c) ϕ; (d) $\overrightarrow{CD}$; (e) $\overline{BC}$

5 Yes, yes

7 When they are distinct non-parallel lines in the same plane; when they are the same line; when they are parallel lines in a plane or skew lines in three dimensions; no

9 (a)

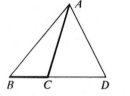

(b) $\overline{BD}$; (c) B; (d) A; (e) ϕ

11 Infinite

13 All

15
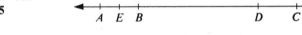

EXERCISE 2B

5 There is no distinction between them.

9 Any segment on $\overleftrightarrow{AD}$ whose endpoints are between A and D; $\overset{\circ}{AB}$, $\overset{\circ}{OB}$, $\overset{\circ}{OC}$, $\overset{\circ}{DC}$, $\overline{BC}$ is the only segment shown.

11 The tops of the legs are always in the same plane; hence, it will not wobble on any surface.

EXERCISE 2C

1 Yes

3 They are independent properties.

     (a)   (b)   (c)   (d)

5 Postulates 2.1 and 2.2

7 Two segments do not yield a closed figure.

9 (a), (c), (e) are closed; (a), (b), (d), (e) are simple; (c), (e) are polygons; (c), (d), (e) are polygonal paths.

11 Given a polygon such that the intersection of the interior points of any two sides is not empty, then the polygon is non-simple.

EXERCISE 2D

1 No

3 (a), (d), (f), (g) separate the plane, and (b) might, depending on the position of the rope; in (f) the letters A, B, D, O, P, Q, R separate the plane.

5 Four faces are required to separate space. The minimum polygon for a face is

258

a triangle. If one triangle is used for a base, three more are needed to bound each edge of the base triangle.

7 4, 6, 8, 9, 0; zero is the only simple closed curve.

EXERCISE 2E

1 (b), (c), (f), (g) are convex; (a) is likely to be; (i) is in general overall shape; the rest are not or are not likely to be convex.

9 No

11

13

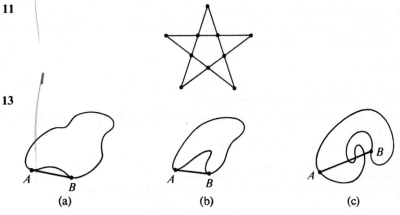

(a) (b) (c)

If a line segment connecting any two points on the curve exists such that $\overline{AB}$ is in part exterior to the figure, then the figure is concave.

EXERCISE 3A

1 Itself

3 $\overline{AC}$

5 $\overline{AB} \ncong \overline{CD}$

7 Two segments congruent to the same segment are congruent to each other. Any segment is congruent to itself.

9 Any segment is congruent to itself.

11 None

13 $\overline{NO}, \overline{OQ}; \overline{MQ}, \overline{NP}; \overline{NM}; \overline{PQ}; \overline{MO}, \overline{OP}$

15 Yes, yes, yes, equivalence

EXERCISE 3B

1 True

3 True

5 False

7 False

9 False

12 $\angle ODC \cong \angle OCD \cong \angle ABD; \angle OCB \cong \angle OBC \cong \angle ADB$

EXERCISE 3C

1 Yes, side-angle-side
3 Not necessarily
5 Yes, side-side-side
7 No; triangles can have proportional sides.
9 Side-side-side
11 $\triangle PMN \cong \triangle NQP$, side-side-side; $\triangle PMR \cong \triangle NQR$, side-side-side; $\triangle MRN \cong \triangle QRP$, side-side-side
13 Yes; triangles are congruent right triangles.
14 Construct congruent triangles using angle-side-angle and measure the corresponding segment on dry land.

EXERCISE 3D

1 Side-angle-side
3 Yes; corresponding angles are congruent.
7 One; the sum of the angles cannot exceed two right angles and there must be three angles.
9 $\angle ABC \cong \angle BCE$; $\angle BAC \cong \angle ECD$
11 No
13 Subtract equals from equals leaving equals.
16 $\angle 7 \cong \angle 8$; $\angle 6 \cong \angle 1$; $\angle 5 \cong \angle 2$; $\angle CAB \cong \angle 7$ and $\angle 8$

EXERCISE 3E

3 Three angles are needed.
5 Yes
7 Exercise 3E(5) involves plane figures; Exercise 3E(6a) requires a fourth-dimensional "flip."

EXERCISE 4A

1 The union of the interior angles is congruent to two straight angles.
5 The use of parallel lines and transversal concepts insure the other angles are right angles.
7 No, need a right angle too.
13 $\overline{AB} \cong \overline{CD}$; $\angle A \cong \angle C$; $\angle AFB \cong \angle CED$; $\therefore \triangle ABF \cong \triangle CED$

EXERCISE 4B

3 Yes; yes
5 Four
7 Twenty
8 No; new vertices are formed such that four edges meet at some vertices.

EXERCISE 4C

1 Equal radii
7 Yes; perpendicular distances from each face are equal.

260

8 Circle or point
9 Infinite number, one
10 Infinite; one
12 Yes
13 No; no

EXERCISE 4D

1 Cube; tetrahedron; cube; cube
3 $F = S + 2$
5 Yes; an oblique section of a rectangular pyramid
7 $E = 2n$
9 Yes
12 No

EXERCISE 4E

1 Yes, yes
3 No
5 Cubes, spheres
7 $\triangle BAO \sim \triangle COD$; $\triangle AOD \sim \triangle BOC$; $\triangle BCA \sim \triangle BCD$; $\triangle BDA \sim \triangle DCA$
9 Should be similar, ignoring distortion
11 Three

EXERCISE 4F

1 (a) Translation, rotation; (c) rotation; (e) rotation; (g) similarity; (i) rotation
3 The plane, a point, a line
5 Yes; let the ratio be 1:1
9 (b)

EXERCISE 4G

1 (a) No symmetry; (c) line symmetry; (e) line and point symmetry; (g) line and point symmetry
3 (a) Yes; (c) for even number of vanes, yes; (e) no; (g) yes (small details excepted)
7 A, B, C, D, E, H, I, K, M, O, T, U, V, W, X, Y all have line symmetry; H, I, N, O, S, X, Z have point symmetry.
9 All
13 TOT, WOW, and others
17 Yes
19 5, 6, 8; for a regular star of n points there are n lines of symmetry.

EXERCISE 5A

1 Equals
3 $\overline{AC}$
11 2, 5, 3, 10, 14, 9, 4

261

EXERCISE 5B

  4 $C \doteq 3.14d$

  6 $2a + 2b$

  8 $2b + 2\sqrt{h^2 + k^2}$, where $h^2 = 2bk - k^2$

EXERCISE 5C

  3 Yes

  5 $180°$ or m(straight angle), π radians

  7 $180°$ or m(straight angle), yes

  9 Yes

 11 In (a), $m(\measuredangle COD) = 50°, m(\measuredangle AOD) = 130°, m(\measuredangle BOC) = 130°$.
      In (c), $m(\measuredangle AEC) = 57°$

EXERCISE 5D

  1 The unit squares counted never completely fill the circle.

  3 8, 14, 25

  9 $5\sqrt{3}$

 11 18

 12 Area is doubled; area is six times larger; area is the same.

 16 C is doubled, A is four times larger.

 18 $25(\pi - 2) \doteq 28.50$ square units

EXERCISE 6A

  1 (a) 240 cubic units; (b) 160 cubic units; (c) 192 cubic units

  3 $7200 + 600\sqrt{11}$

  5 35,640 cubic inches; approximately 20 cubic feet

  7 $V = A_b h$ is general formula; if s is the measure of one side of the base and d is
      the measure of the distance from a vertex to the center, then
      $A_b = 5\left(\frac{1}{2}s\sqrt{d^2 - s^2}\right) = 5s\sqrt{d^2 - s^2}/2$.

  9 10,725 cubic feet

 11 $\frac{2}{3}$ cubic yard

 13 $8\sqrt[3]{2}$; $4\sqrt[3]{2}$; $7.2\sqrt[3]{2}$

EXERCISE 6B

  1 (a) 250π; (b) 500π; (c) 250π; (d) $432\sqrt{3}\pi$

  3 $4800/\pi$ cubic feet

  5 17 feet (approximately)

  7 $93\pi + 192$ cubic inches

  9 2025π cubic inches

 11 277.3π cubic feet

 13 $4500/3\pi^2$; $62,500/\pi^2$; $500,000/3\pi^2$

 15 4 square units; 6 units; 24 cubic units; 4 square units

EXERCISE 6C

1 (a) 48 cubic units; (b) 30π cubic units; (c) 39π cubic units; (d) 32 cubic units; (e) $16\sqrt{231}$ cubic units
3 364π cubic inches; 4.9 gallons
5 Answer is "no" if sand must be level with top, "yes" if it can be piled.
7 Pots contain $490,000\pi$ cubic inches, or approximately 33 cubic yards, of dirt.
9 $375\pi/96$ cubic inches

EXERCISE 6D

1 In a cube the length $= e$, the width $= e$, and the height $= e$; hence, using Formula 6.12, $S = 2lw + 2wh + 2lh = 2e^2 + 2e^2 + 2e^2 = 6e^2$.
5 $S = A_b + sP$
7 $S = (e/4),(6s + \sqrt{3}e)$, where e is an edge and s is the slant height
9 $S = \pi(Rs + rs + r^2 + R^2)$
11 (a) 138 square units; (b) $16 + 24\sqrt{5}$ square units; (c) $\frac{64}{3}\sqrt{65}\pi$ square units; (d) 40.5π square units; (e) 196π square units; (f) $288 + 8\sqrt{3}$ square units
13 Need slightly more than 5 gallons; if only gallons can be purchased, cost $=$ $60.
15 The volume of each pyramid approaches $\frac{1}{3}A_b r$, where r is the radius of the sphere. The sum of all the pyramids is $\frac{1}{3}A_{b_1}r + \frac{1}{3}A_{b_2}r + \cdots + \frac{1}{3}A_{b_n}r = \frac{1}{3}r(A_{b_1} + A_{b_2} + \cdots + A_{b_n}) = \frac{1}{3}rS = (\frac{1}{3}r)(4\pi r^2) = \frac{4}{3}\pi r^3$.

EXERCISE 7A

1 (a) 3.28 feet, 1.09 yards; (b) 2.113411 pints, 1.056705 quarts, 60.98 cubic inches
3 Exercise 7A(2) was easier.
5 (a) 1 meter $=$ 1.09 yards, 1 yard $=$ 0.927 meter
    (c) 1.63 kilometers $=$ 1 mile, 1 kilometer $=$ 0.62 miles
    (e) 1 metric ton $=$ 1.10 short tons, 1 short ton $=$ 0.907 metric ton
    (g) 1 liter $=$ 1.057 quarts, 1 quart $=$ 0.947 liter
    (i) 1 gallon $=$ 3.788 liters, 1 liter $=$ 0.264 gallon
7 1500 meters $=$ 1635 yards, 1635 yards $<$ 1760 yards $=$ 1 mile
11 4.8 cubic centimeters $=$ 1 teaspoon

EXERCISE 7B

1 Yes; evaluation rate of exchange varies; U.S. monetary system
3 $F = 32 + \frac{9}{5}C$; (a) $-58°F$; (c) $248°F$; (e) $31\frac{1}{9}°C$.

EXERCISE 7C

3 Hundredths
5 Different precision, different standards
7 (a) is to the nearest tenth; (b), (c), (g), (h), (i) to the nearest hundredth.
9 Approximate

263

EXERCISE 7D

1   (a) 3;   (c) 3;   (e) 6;   (g) 3;   (i) 4;   (k) 6
3   Three significant digits: (a) 9840;   (c) 0.00173;   (e) 2700
    Two significant digits: (a) 9800;   (c) 0.017;   (e) 2700
    One significant digit: (a) 10,000;   (c) 0.002;   (e) 3000
5   Hundredths: (a) 472.14;   (c) 10.10;   (e) 0.00
    Units: (a) 472;   (c) 10;   (e) 0
    Tens: (a) 470;   (c) 10;   (e) 0
    Hundreds: (a) 500;   (c) 0;   (e) 0
7   4(e) above
9   No

EXERCISE 7E

1   $\frac{1}{500}$, $\frac{1}{40}$, 95 miles, 1 inch
3   113 to 127 feet
5   13.5 feet
7   (a) 16.8 feet; (c) 196 tons

EXERCISE 7F

1   (a) 0.105;   (c) 0.01
3   6.57 acres;   6.60 acres
5   \$298–\$428
9   (a) 24,617.2;   (c) 276,827.31

EXERCISE 8A

1   (a) 5;   (b) 3;   (c) -5
3   (a) -14;   (b) -17;   (c) 14
5   No
7   $m_d(\overline{AB}) = x_2 - x_1$; $m_d(\overline{BC}) = x_3 - x_2$; $m_d(\overline{AB}) + m_d(\overline{BC}) =$
    $x_2 - x_1 + x_3 - x_2 = -x_1 + x_3 = x_3 - x_1 = m_d(\overline{AC})$
9   $(-4, 3), (-4, -3), (4, -3)$
11 Zero
13 One set of vertices to complete the square is $(-3, -3)$ and $(1, -3)$; more than
    one set exists.
15 Yes

EXERCISE 8B

1   (a) $\sqrt{41}$;   (b) $5\sqrt{2}$;   (c) $2\sqrt{29}$;   (d) $\sqrt{157}$
3   (a) $(2\frac{1}{2}, 5)$;   (b) $(1\frac{1}{2}, 1\frac{1}{2})$;   (c) $(0, 2)$;   (d) $(\frac{1}{2}, 1)$
5   Yes
7   No
9   All segments are equal and right angles are formed; use distance formula to
    verify.
11 They all fall along one straight line; $x + y = 4$.

264

13 $(6, 1), (-2, 1), (0, 1)$ are several points on the line; no matter what value is chosen for x, $y = 1$.

15 Yes; $y = x + 2$

EXERCISE 8C

1 (a) Not a straight line; (b) straight line; (c) straight line; (d) straight line

3 (a) Not a parallelogram; (b) parallelogram; (c) not a parallelogram; (d) not a parallelogram

5 -4; it is the same; the line which passes through the points (x_1, y_1) and (x_2, y_2) has the same slope as the line that passes through (kx_1, ky_1) and (kx_2, ky_2), where k is any non-zero constant.

7 $\frac{5}{3}$ and $-\frac{5}{3}$

EXERCISE 8D

1 (a) Not parallel; (b) not parallel; (c) same line; (d) parallel

3 The ratio $(y_2-y_1)/(x_2-x_1)$ has a denominator of 0.

5 $m(\overline{OB}) = \sqrt{3^2 + 3^2} = \sqrt{18}$; $m(\overline{OA}) = \sqrt{3^2 + 3^2} = \sqrt{18}$; $m(\overline{AB}) = \sqrt{6^2} = \sqrt{36}$; $m(\overline{OB})^2 + m(\overline{OA})^2 = 18 + 18 = 36 = m(\overline{AB})^2$

7 3

9 Yes; yes; yes

EXERCISE 8E

1 Area $= \frac{1}{2}(x_2y_1 + x_3y_2 + x_1y_3 - x_1y_2 - x_2y_3 - x_3y_1)$

3 $51\frac{1}{2}$

5 $4\frac{1}{2}$

7

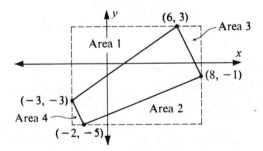

Area of entire area $= (11)(8) = 88$ square units

Area 1 $= \frac{1}{2}(6)(9) = 27$ square units

Area 2 $= \frac{1}{2}(10)(4) = 20$ square units

Area 3 $= \frac{1}{2}(2)(4) = 4$ square units

Area 4 $= \frac{1}{2}(1)(2) = 1$ square unit

Area of quadrilateral is 36 square units.

265

EXERCISE 8F

1

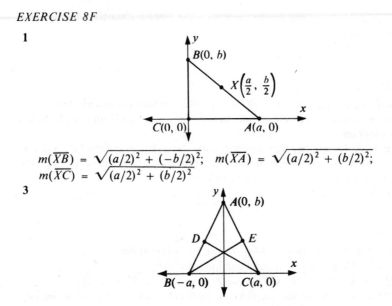

$m(\overline{XB}) = \sqrt{(a/2)^2 + (-b/2)^2}; \quad m(\overline{XA}) = \sqrt{(a/2)^2 + (b/2)^2};$
$m(\overline{XC}) = \sqrt{(a/2)^2 + (b/2)^2}$

3

D and E have coordinates $(-a/2, b/2)$ and $(a/2, b/2)$;
$m(\overline{DC}) = \sqrt{(a/2)^2 + (b/2)^2}$ and $m(\overline{BE}) = \sqrt{(a/2)^2 + (b/2)^2}$.

5 If each side of a rectangle has a measure of $2A$, then the line joining the mid-points of the opposite in each case has a measure of A; hence, the new triangle is equilateral

7

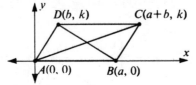

$m(\overline{AD}) = \sqrt{b^2 + k^2}$, therefore $m(\overline{CB}) = \sqrt{b^2 + k^2}$
$m(\overline{AB}) = a$, therefore $m(\overline{DC}) = a$
$m(\overline{CA}) = \sqrt{(a + b)^2 + k^2}$ and $m(\overline{DB}) = \sqrt{(b - a)^2 + k^2}$
then $m(\overline{AD})^2 + m(\overline{CB})^2 + m(\overline{AB})^2 + m(\overline{DC})^2 =$
$$b^2 + k^2 + b^2 + k^2 + a^2 + a^2 = 2(a^2 + b^2 + k^2)$$
and $m(\overline{DB})^2 + m(\overline{AC})^2 = (a + b)^2 + k^2 + (b - a)^2 + k^2 =$
$$a^2 + 2ab + b^2 + k^2 + b^2 - 2ab + a^2 + k^2 = 2(a^2 + b^2 + k^2)$$

EXERCISE 8G

1 (a) $y = 2x - 5$; (b) $y = \frac{1}{2}x + 1$; (c) $y = -\frac{1}{2}x + 5$; (d) $y = -4x + 19$
3 An infinite number; one

266

7 (a)   (b)   (c)

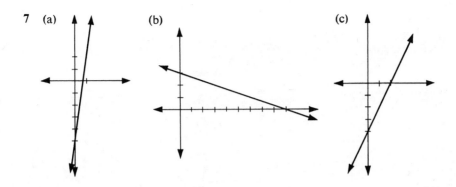

EXERCISE 8H

1 $x^2 + y^2 = r^2$
3 Yes
5 $x^2 + y^2 - 4y = 21$
7 $x^2 + y^2 - 4x - 6y = 23$
9 $x^2 + y^2 = 2$

INDEX

269

271